# Collective Models of the Nucleus

# PURE AND APPLIED PHYSICS

## A SERIES OF MONOGRAPHS AND TEXTBOOKS

### CONSULTING EDITORS



Volume 1. F. H. FIELD and J. L. FRANKLIN, Electron Impact Phenomena and the Properties of Gaseous Ions.
Volume 2. H. KOPFERMANN, Nuclear Moments, English Version Prepared from the Second German Edition by E. E. SCHNEIDER.
Volume 3. WALTER E. THIRRING, Principles of Quantum Electrodynamics. Translated from the German by J. BERNSTEIN. With Corrections and Additions by WALTER E. THIRRING.
Volume 4. U. FANO and G. RACAH, Irreducible Tensorial Sets.
Volume 5. E. P. WIGNER, Group Theory and Its Application to the Quantum Mechanics of Atomic Spectra. Expanded and Improved Edition. Translated from the German by J. J. GRIFFIN.
Volume 6. J. IRVING and N. MULLINEUX, Mathematics in Physics and Engineering.
Volume 7. KARL F. HERZFELD and THEODORE A. LITOVITZ, Absorption and Dispersion of Ultrasonic Waves.
Volume 8. LEON BRILLOUIN, Wave Propagation and Group Velocity.
Volume 9. FAY AJZENBERG-SELOVE (ed.), Nuclear Spectroscopy. Parts A and B.
Volume 10. D. R. BATES (ed.), Quantum Theory. In three volumes.
Volume 11. D. J. THOULESS, The Quantum Mechanics of Many-Body Systems.
Volume 12. W. S. C. WILLIAMS, An Introduction to Elementary Particles.
Volume 13. D. R. BATES (ed.), Atomic and Molecular Processes.
Volume 14. AMOS DE-SHALIT and IGAL TALMI, Nuclear Shell Theory.
Volume 15. WALTER H. BARKAS. Nuclear Research Emulsions. Part I.
Nuclear Research Emulsions. Part II. *In preparation*
Volume 16. JOSEPH CALLAWAY, Energy Band Theory.
Volume 17. JOHN M. BLATT, Theory of Superconductivity.
Volume 18. F. A. KAEMPFFER, Concepts in Quantum Mechanics.
Volume 19. R. E. BURGESS (ed.), Fluctuation Phenomena in Solids.
Volume 20. J. M. DANIELS, Oriented Nuclei: Polarized Targets and Beams.
Volume 21. R. H. HUDDLESTONE and S. L. LEONARD (eds.), Plasma Diagnostic Techniques.
Volume 22. AMNON KATZ, Classical Mechanics, Quantum Mechanics, Field Theory.
Volume 23. WARREN P. MASON, Crystal Physics of Interaction Processes.
Volume 24. F. A. BEREZIN, The Method of Second Quantization
Volume 25. E. H. S. BURHOP (ed.), High Energy Physics. In three volumes.
Volume 26. L. S. RODBERG and R. M. THALER, Introduction to the Quantum Theory of Scattering
Volume 27. R. P. SHUTT (ed.), Bubble and Spark Chambers. In two volumes.
Volume 28. GEOFFREY V. MARR, Photoionization Processes in Gases
Volume 29. J. P. DAVIDSON, Collective Models of the Nucleus

*In preparation*

H. S. GREEN and R. B. LEIPNIK, The Foundations of Magnetohydrodynamics and Plasma Physics

# COLLECTIVE MODELS OF THE NUCLEUS

J. P. Davidson

DEPARTMENT OF PHYSICS
UNIVERSITY OF KANSAS
LAWRENCE, KANSAS

1968

ACADEMIC PRESS New York and London

ACADEMIC PRESS INC.
111 Fifth Avenue New York, New York 10003

*United Kingdom Edition published by*
ACADEMIC PRESS INC. (LONDON) LTD.
Berkeley Square House, London W. 1

LIBRARY OF CONGRESS CATALOG CARD NUMBER: 68-18665

PRINTED IN THE UNITED STATES OF AMERICA

*This book is dedicated to*
*the memory of my father,*
*John Pirnie Davidson*

# Preface

The theory of nuclear structure can be more or less divided into two parts having common origins in time but becoming more intertwined as our knowledge of the nucleus has become deeper. The shell model, on one hand, singles out one or a few nucleons and attempts through, by now, rather detailed calculations with specific potentials and residual interactions to explain not only general features displayed by nuclei over fairly wide regions but also to explain and predict properties of individual nuclei. On the other hand, the collective model endows a quantized fluid with certain properties and subjects it to certain boundary conditions in order to predict those gross characteristics which depend upon most, if not all, of the nucleons in a system acting in concert. In order to explain certain other properties, especially those having to do with a single nucleon or two superimposed upon a more collective background, one must amalgamate these extremes into a unified model.

In preparing a summary of these last two topics for *Reviews of Modern Physics* I found that while there existed several monographs on the shell model, in particular those by Feenberg, Mayer and Jensen and, more recently, de Shalit and Talmi, there existed no similar volume on the collective model. This seemed strange and quite a pity since collective models have played a very important role in codifying the vast amounts of experimental information flowing from physics laboratories all over the world and have also formed a platform from which numerous investigations of the nuclear many-body problem have proceeded. That such a monograph as this would be useful was obvious even though the question of content and emphasis was not. One could either bring together and detail the various collective models from a phenomenological point of view, or one could take the microscopic trail and display and explain the collective models from an underlying and perhaps more fundamental viewpoint, or both. I have chosen the former path not only because of its usefulness to theorist and experimentalist but also because there exist two books, one by Lane and the other by Brown, which quite adequately cover our current state of knowledge from the microscopic point of view.

I have tried to bring together ideas and results of calculations, many "well known" but not very accessible, and combine them into an interdependent discussion on those aspects of nuclear structure which rely for their interpretation on various collective modes of motion. This has resulted in the omission of a discussion of particle reactions such as stripping and multinucleon transfer reactions for which particle aspects are more important than collective ones. Similar reasoning has led to including photonuclear reactions where collective effects would seem to play the dominant role. The volume should thus be useful to both experimental and theoretical physicists. The former will find it helpful in codifying the results of current experiments and suggesting new ones. The latter may be able to use it as a springboard to new extensions of these current theories as well as the outline of a set of model characteristics to be explained by detailed microscopic calculations.

The organization of the book is from old to new, even to odd, static to dynamic. The short first chapter is basically historical in nature and serves only as a skippable introduction. While devoting such space to an historical introduction may not be out of place in a monograph such as this it is somewhat out of fashion. However, a reexamination of early related ideas has an important place in physics even if it is only pedagogic. I cannot help but feel that Niels Bohr's first paper, for which he received the Gold Medal of the Royal Danish Scientific Society, had an important impact on the later development of nuclear theory even though the classical ideas contained within it lay dormant for upwards of thirty years.

Following the introduction is a chapter devoted to some of the physics of a liquid drop. The ideas developed there form the basis of the various vibrational models as well as the method by which vibrational degrees of freedom are added to rotational models. The following three chapters are each devoted to a different class of nuclei, even-even, odd-$A$, and odd-odd, and each chapter builds upon the earlier ones. In each of these model chapters I have tried first to display those characteristics which follow from the mere existence of a deformed shape and are essentially independent of the specific details of the various models. Following these in turn are chapters involving the various types of nuclear transitions: gamma, alpha and beta, photonuclear, and finally those involving mu-mesic atoms. Only this last needs any justification and I have included it since the mu-meson seems to present a new and powerful tool for the investigation of collective nuclear phenomena. To date discussions of the interaction of mu-mesons with collective nuclear modes have dealt with only the simplest of these latter. These predictions have been amply confirmed. However, with solid state detectors and the coming meson "factories," experimental physicists will have the tools for much more sophisticated investigations and I have tried at least to suggest how one might proceed from the current calculations. I have placed in Appendices certain useful mathematical details which are important for the carrying out of calculations.

A word about references and the numbering of equations is in order. The reference list is not exhaustive and the references listed are those which give the major support of an idea or calculation. Since the Bibliography is ordered into a section of books, one of review articles, and lastly one of papers, I feel that those wanting a complete list of contributors to an idea should consult the most recent review article. The same is true of experimental support. Most measured values, like skirt lengths, change with time (the most amusing example of the former is the value of the velocity of light) and as a theorist I would rather not get involved with whose slip is showing. For this reason I have included almost no tables of experimental data. The data are important but the numbers are ephemeral. Again this is the role of the review articles. As for the equations those which are referred to are numbered consecutively within a chapter. When referred to within the chapter only the arabic numeral is given; when referred to from without the chapter the numeral of the chapter is included.

Finally, I should like to acknowledge the help I have had from many sources. Some of the work described herein is that of my students, Drs. P. O. Lipas, A. K. Rafiqullah, S. A. Williams, B. E. Chi, M. G.Davidson, and J. R. Roesser. All have helped me to better understand these ideas. In particular, the tables and drawings of the Nilsson orbitals in Appendix D are due to Dr. Chi. I also wish to thank Dr. S. A. Williams and Professor Paul Goldhammer for suggestions made as a result of reading the entire manuscript. The errors remaining are mine and not theirs. And last but certainly not least many thanks are due to an old friend, Professor W. A. McKinley at Rensselaer Polytechnic Institute for many years of valuable and informative discussions.

*March 1968* J. P. DAVIDSON

# Contents

# Collective Models of the Nucleus

CHAPTER 1

# Introduction

The first suggestion that collective phenomena might occur in atomic nuclei antedates both the discovery of the neutron by Chadwick in 1932 and the discovery of the *magic numbers* and the use of the first shell model ideas by Elsasser [86] in 1934. In 1930 Thibaud [152] pointed out that the recent, fine-structure, alpha-decay measurements by Rosenblum [143] on $Bi^{212}$ might be explained if the daughter nucleus $Tl^{208}$ could be considered a system in rotation. Thibaud used, in his discussion, both the eigenvalue relations for axially symmetric and spherical rotors. This work is now only of historical interest since thallium is not in what is now considered to be one of the deformed regions and does not possess a low energy rotational structure. Considerably later, Niels Bohr and Kalckar in discussing the energy level distribution in heavy atomic nuclei pointed out that the level density, as a function of the mass number $A$, behaved more or less like that of the surface vibration of a quantum fluid [59] (they also considered the elastic vibrations of a quantum solid). Some of the details of the problem of surface waves on a spherical volume of an irrotational fluid were worked out by Bethe, who also discussed longitudinal volume waves in a compressible fluid [25]. Shortly thereafter, these ideas of the vibrations of a liquid drop were to be so important in understanding the newly discovered phenomenon of nuclear fission.

Although fission was not predicted by nuclear theorists, they were quick to realize its implications and to use the liquid-drop model to explain it. Feenberg [88], Frenkel [93], N. Bohr and Wheeler [60], and von Weizsäcker [156] all studied the relation between shape and stability of a uniformly charged but deformed liquid drop and how the total energy changed as a function of deformation from the spherical. Feenberg's note was basically concerned with demonstrating the stability of nuclei against spontaneous fission. Bohr and Wheeler, as well as Frenkel, investigated the theory of the process in quite some detail. Indeed Frenkel was somewhat more perspicacious when he noted that "It seems possible that if the ratio $E/U$ is not too large, a stable nonspherical form may be assumed.... If this view should prove correct, the electrical quadrupole moments of the heavier nuclei should be abnormally large" [93]. At this time at least three nuclei—$Eu^{151}$, $Eu^{153}$, and $Lu^{175}$—were known to have rather large quadrupole moments. Casimer [4] had shown some three years earlier in an analysis of the deviations of the optical spectra from the interval rule that these quadrupole moments were about 1.5, 3.2, and 5.5 barns, respectively.

Instability in the liquid-drop model arises when the volume energy associated with the coulomb forces of repulsion of the protons exceeds the surface energy of contraction. By applying a slight, and volume conserving, distortion to the volume of the drop, it is clear that the surface, and thus the surface energy, will increase while the coulomb energy will decrease. If the circumstances are such that the coulomb energy decreases faster than the surface energy increases, then an instability exists which can lead to fission. Many years ago Lord Rayleigh investigated a similar classical problem in which he studied the stability of charged liquid drops in a jet [151]. Refinements in the theory were supplied by Niels Bohr in his first published work [58] which was done even before Rutherford postulated his nuclear model of the atom [145]. (Similar investigations have been carried out concerning the stability of a rotating mass of gas. The boundary condition is different but the coulomb and gravitational potentials are identical in form even if opposite in sign. These matters are of astrophysical interest but the mathematical details are easily applied to the problem being considered. For a detailed mathematical treatment see Lyttleton [13].)

In order to lay the groundwork for what is to follow we reproduce here in a somewhat more general way the discussion of Feenberg [88]. Consider deforming a sphere of radius $R_0$, while preserving its volume, into an ellipsoid given by

$$\lambda x^2 + \varkappa y^2 + z^2/\lambda\varkappa = R_0^2. \tag{1}$$

For small deformations the eccentricities may be expressed in terms of the parameters $\varkappa$ and $\lambda$ by

$$e_1 = (a - c)/R_0 \cong 2(1 - \sqrt{\lambda}) + (1 - \sqrt{\varkappa})$$
$$e_2 = (b - c)/R_0 \cong (1 - \sqrt{\lambda}) + 2(1 - \sqrt{\varkappa})$$

and we have labeled the semiaxes so that $a > b > c$. Calling $E_C$ and $E_S$, respectively, the coulomb and surface energies of the deformed system and $E_C^0$ and $E_S^0$, respectively, the corresponding energies of the original sphere, then since the surface energy is simply proportional to the surface area, we work with this. The surface area of an ellipsoid in terms of the semiaxes is given by [3]

$$S = 2\pi c^2 + \frac{2\pi b}{\sqrt{a^2 - c^2}} [(a^2 - c^2)\, E(\theta) + c^2\theta] \tag{2}$$

where $E(\theta)$ is an elliptic integral of the second kind defined by

$$E(\theta) = \int_0^\theta dn^2 u \, du \tag{3}$$

while $\theta$ is defined by

$$\frac{a^2 - c^2}{a^2} = \mathrm{sn}^2(\theta,\ k),$$

where the modulus is defined by

$$k^2 = \frac{b^2 - c^2}{a^2 - c^2}$$

and $\mathrm{sn}\, u$, $\mathrm{cn}\, u$, and $\mathrm{dn}\, u$ are the usual elliptic functions. For our purposes it is sufficient to expand $\mathrm{sn}^{-1} u$ and $\mathrm{dn}^2 u$ in a power series then integrate Eq. (3) term by term and substitute in (2). This process yields, after a bit of algebra, the surface energy of an ellipsoidally shaped nucleus

$$E_S = E_S^0[1 + (8/45)\,(e_1^2 - e_1 e_2 + e_2^2) + \cdots]. \tag{4}$$

The coulomb energy for a uniformly charged ellipsoid can be given in closed form since it is similar to the gravitational energy of an ellipsoid mass. Following Lyttleton [13] we may write

$$E_C = -\frac{3}{10}\left(\frac{Ze}{R_0^{\,3}}\right)^2 a^2b^2c^2 \int_0^\infty \frac{d\Gamma}{\Delta}$$

$$= E_C{}^0 \frac{a^2b^2c^2}{2\,R_0^{\,5}} \int_0^\infty \frac{d\Gamma}{\Delta}, \tag{5a}$$

where

$$\Delta^2 = (a^2 + \Gamma)(b^2 + \Gamma)(c^2 + \Gamma)$$

and $\Gamma$ is the "radial-like" ellipsoidal coordinate. The integral in (5a) can be expressed in terms of the inverse elliptic functions

$$E_C = E_C{}^0 \frac{\sqrt[3]{abc}}{\sqrt{a^2 - c^2}} \operatorname{sn}^{-1} \sqrt{\frac{a^2 - c^2}{a^2}}, \tag{5b}$$

where the modulus is now

$$k^2 = \frac{a^2 - b^2}{a^2 - c^2}.$$

On again expanding $\operatorname{sn}^{-1}u$ in a power series of its argument, the relation (5b) becomes

$$E_C = E_C{}^0[1 - (4/45)(e_1{}^2 - e_1e_2 + e_2{}^2) + \cdots] \tag{6}$$

Expressions (4) and (6) have the expected behavior for small deformations. The question of stability is resolved by studying the change in total energy, coulomb plus surface, which is simply

$$\Delta E_d = \frac{4\,E_S{}^0}{45}\left(2 - \frac{E_C{}^0}{E_S{}^0}\right)(e_1{}^2 - e_1e_2 + e_2{}^2). \tag{7}$$

Stability against fission exists so long as $E_C{}^0/E_S{}^0 < 2$; otherwise the system will undergo spontaneous fission. This relation was derived in the early papers mentioned before. (Frenkel used $E$ and $U$ in place of $E_C{}^0$ and $E_S{}^0$ respectively.)

Now if we treat the eccentricities in Eq. (7) as variational parameters we see immediately that the most stable shape is the spherical one. From just such arguments as these it was concluded that the low energy shapes of even-even nuclei and the even-even cores of other nuclei were spherical. If this were true then, of course, the low energy level structure of nuclei would not be rotational in nature. (After Thibaud's original suggestion, the idea of explaining nuclear level structure within the context of a deformed quantum-mechanical rotor was put forward from time to time. However, the rather meager knowledge of spins and parities of excited nuclear states precluded any experimental justification for models such as these. This was true even as late as about 1950 as can be seen from examining the compendium of nuclear structure data of the time [39]. Not until about 1954 was a nucleus shown to have several levels with spacings suggestive of those of a deformed rotor [131].)

From the point of view of the liquid-drop model it would be necessary to add a term linear in the eccentricities to Eq. (7) in order to obtain static, deformed shapes. The necessity for a term like this was not seen for some ten years. Following the end of World War II a great deal of effort went into measuring the magnetic dipole and electric quadrupole moments of nuclei. During this period, when theory on the shell model was rapidly being developed, it became clear that although the electric quadrupole moments displayed shell effects—being small and changing sign in passing through a shell edge and large in passing through the middle [104]—they were far too large in the rare earth regions to be accounted for by one or by even a few protons moving in individual orbitals as required by the shell model. It was suggested that these very large moments were indicative of nuclear polarization [154]. Almost immediately Rainwater showed that a single particle moving in a deformed oscillator well might have an energy lower than when moving in a similar spherical well [139]. It turned out that this change in energy was proportional to the first power of the eccentricity so that static deformations of the liquid drop were possible. This process then unified the shell model and the liquid-drop model.

Although Rainwater used an oscillator potential, it is probably simpler to use a square well of finite depth since only the numerical factors are different. We follow a method given by Feenberg and Hammack [89] but generalize it to the case of an ellipsoidal deformation. Consider a particle moving in a deformed well of depth $D$ defined by

$$t^2 = \lambda x^2 + \varkappa y^2 + z^2/\lambda\varkappa,$$

the potential being given by

$$\begin{aligned} V(t) &= -D, \qquad t < R_0 \\ &= 0, \qquad t \geq R_0. \end{aligned}$$

The change in energy due to the deformation of the well from the spherical shape is just

$$\Delta E_p = E_p(e_1, e_2) - E_p(0) = \int |\psi_0(x, y, z)|^2 [V(t) - V(0)] \, dx \, dy \, dz$$

$$= \int V(r) \left[ \left| \psi_0 \left( \frac{x}{\sqrt{\lambda}}, \frac{y}{\sqrt{\varkappa}}, z\sqrt{\varkappa\lambda} \right) \right|^2 - |\psi_0(x, y, z)|^2 \right] dx \, dy \, dz.$$

To proceed, expand the first term in square brackets in a Taylor's series about the unperturbed solution, use the divergence theorem, and finally transform to spherical polar coordinates.

$$\begin{aligned}\Delta E_p &= -\frac{R_0{}^3 D}{3} \left\{ \sqrt{\frac{6\pi}{5}} (e_1 - e_2) \int_s [Y_{22}(\theta, \varphi) + Y_{2-2}(\theta, \varphi)] \right. \\ &\quad \times |\psi_0(R_0, \theta, \varphi)|^2 \, d\Omega \\ &\quad \left. - 2\sqrt{\frac{\pi}{5}} (e_1 + e_2) \int_s Y_{20}(\theta, \varphi) \, |\psi_0(R_0, \theta, \varphi)|^2 \, d\Omega \right\} \\ &= DR_0{}^3 \mathscr{R}_{nl}^2 \frac{l(l+1) - 3m^2}{3(2l-1)(2l+3)} (e_1 + e_2). \end{aligned} \tag{8}$$

Here we have used for the unperturbed solutions

$$\psi_0(r, \theta, \varphi) = \mathscr{R}_{nl}(r) \, Y_{lm}(\theta, \varphi).$$

For an odd-$A$ nucleus the total energy is the sum of the contributions from the liquid drop and the particle. Thus adding Eqs. (7) and (8) we obtain

$$\begin{aligned}\Delta E &= \Delta E_d + \Delta E_p \\ &= A(e_1{}^2 - e_1 e_2 + e_2{}^2) + B(e_1 + e_2).\end{aligned}$$

Treating the eccentricities as variational parameters shows that the minimum in energy exists for a deformed but symmetric shape given by

$$e_1 = e_2 = -B/A.$$

This then in a more general way is the sense of Rainwater's calculation.

This type of calculation should not be used to support the notion that deformed nuclei must possess an axis of symmetry. Even less support for such an idea is available from the classical theory of rotating liquid masses a subject which, as suggested before, has been thoroughly studied. The results of these calculations show that starting with a spherical shape, the system deforms into a spheroid and then proceeds to distort through a family of figurcs known as the MacLaurin spheroids. At some point this family joins onto the family of Jacobi ellipsoids and the line of stability passes along them, terminating in the pear-shaped figures [13]. The quantum-mechanical analog will not contain these identical features but the results could well be similar and it has been pointed out that the boundary conditions are such that the nuclear system in rotation cannot be spheroidal [50]. This matter can only be resolved outside the domain of these phenomenological models using the Hartree-Fock type of many particle calculations with carefully chosen residual forces of the type done by Baranger and Kumar [47].

The initial application of these ideas to some of the problems of nuclear structure involved calculating the deviations of the measured values of the magnetic dipole moment from the predictions of the single-particle shell model. This was done by using the vibrational model in which surface quanta of angular momentum two were coupled to the angular momentum of the core. [91] A similar investigation was also carried out making use of a rotational model [55] and in both cases the calculated values were in closer agreement with experiment than were those of the single-particle shell model.

These ideas have been developed and extended very rapidly within the last ten to fifteen years and have had an immense impact upon the way in which vast amounts of experimental information has been organized. In particular, we know now that these deformed nuclear models are useful in those regions fairly far from closed shells wherein a large fraction of the nucleons are outside a shell and thus give rise to many coupling possibilities. In fact, we are dealing with a variation of the $j$–$j$ coupling shell model which works so well near the shell edges. The particles do not move within a spherically symmetric potential but within a deformed one which, because of the mixing of $j$ values, takes into account a large amount of the particle correlations left out by the simpler model. This reason, perhaps more than any other, explains why the quite simple models that we are about to discuss work so well.

CHAPTER 2

# Some Physics of a Liquid Drop

Since the collective models with which we shall deal are all derived from the union of the liquid-drop model and the shell model, we shall begin by developing some of the classical theory of a liquid drop, after which the quantization of the system can be done by the usual methods.

Systems of small distortion are most easily investigated by expanding the surface of the drop in spherical harmonics using either a system fixed in the laboratory with coordinates $(r, \theta, \varphi)$ or a system fixed in the drop—the body-fixed coordinate system $(r', \theta', \varphi')$. These two coordinate systems are connected by an Euler-angle transformation, the Euler angles being designated $(\theta_1, \theta_2, \theta_3)$. As is usual with such transformations, the form of the various equations to be developed, but not their content, depends upon which rotations are used and their order in going between laboratory and body-fixed systems. In order to avoid confusion we shall use Rose's convention [15] for these angles as well as the spherical harmonic phase convention of Condon and Shortley [5]. (These matters are reviewed in Appendix A.) Thus the surface can be expressed as

$$R^L(\theta, \varphi) = R_0\Big[\alpha_0{}^* + \sum_{\substack{\lambda>1 \\ \mu}} \alpha^*_{\lambda\mu} Y_{\lambda\mu}(\theta, \varphi)\Big] \tag{1}$$

while in the body-fixed system

$$R^B(\theta', \varphi') = R_0\Big[a_0{}^* + \sum_{\substack{\lambda>1 \\ \mu}} a^*_{\lambda\mu} Y_{\lambda\mu}(\theta', \varphi')\Big]. \tag{2}$$

Since the liquid drop is assumed incompressible, $R_0$ is the radius of the undistorted spherical surface. The constancy of the nuclear volume can be used to evaluate $\alpha_0$ which through third order is

$$\alpha_0 = 1 - \frac{1}{12\pi}\left(3 \sum_{\mu} | \alpha_{\lambda\mu} |^2 + T_\lambda\right)$$

where

$$T_\lambda = [(2\lambda + 1)/4\pi]^{1/2}\, C(\lambda\lambda\lambda; 000) \sum_{\mu\mu'} \alpha^*_{\lambda\mu}\alpha_{\lambda\mu'}\alpha_{\lambda\mu-\mu'} \times C(\lambda\lambda\lambda; \mu - \mu', \mu', \mu). \tag{2a}$$

Since these are first-order theories it is usually adequate to set $\alpha_0$ (or $a_0$) equal to one. In these equations the terms for $\lambda = 1$ have been removed by setting $\alpha_{1\mu} \equiv 0$ since they represent the displacement of the center of mass (Appendix B).

The expansion coefficients of Eqs. (1) and (2) are related by

$$\alpha_{\lambda,\mu} = \sum_{\nu=-\lambda}^{\lambda} D^{\lambda*}_{\mu\nu}(\theta_i) a_{\lambda,\nu}. \tag{3}$$

Furthermore, the condition that the nuclear surface is real imposes the requirement that

$$a_{\lambda,-\mu} = (-1)^{\mu}\, a^*_{\lambda,\mu}$$

and a similar condition for the $\alpha_{\lambda,\nu}$. For a $\lambda$th-order surface there are $(2\lambda + 1)$ independent parameters. Three of these specify the orientation of the system; the others $(2\lambda - 2)$ are associated with the shape.

At this point it is necessary to postulate the nature of the fluid motion within the drop. The simplest postulate is irrotational flow and presumably it will be associated with the lowest energy states of the system. (One should not expect this postulate of irrotational flow to be more than a general guide to investigating nuclear structure inasmuch as the fluid description in condensed matter assumes the particle mean free path to be very small compared with the dimensions of the system. In nuclei the Pauli principle insures that the nucleon mean free path is of the order of the dimensions of the nucleus. An alternate simple model is one of rigid rotation [33]. The empirical evidence indicates a situation between these extremes.) If the flow is irrotational then the velocity is derivable from a potential function which, because of the further assumption that the nuclear density is a constant, is a solution of Laplace's equation. This velocity potential, for small surface distortions about a spherical shape is [151]

$$\chi(r) = -R_0^2 \sum_{\lambda,\mu} \frac{\alpha^*_{\lambda,\mu}}{\lambda} \left(\frac{r}{R_0}\right)^\lambda Y_{\lambda,\mu}(\theta, \varphi). \tag{4}$$

The kinetic energy for such motion is then

$$T = \frac{1}{2} \int_V \varrho_0 \mathbf{v}^2 \, d\tau = \frac{\varrho_0}{2} \int |\nabla \chi|^2 \, d\tau$$

$$= \frac{\varrho_0}{2} \int_S \chi^* \nabla \chi \cdot d\mathbf{S}, \tag{5}$$

where $V$ is the nuclear volume bounded by the surface $S$ whose outward normal is $\mathbf{S}$. Now from (4)

$$\chi^* \nabla \chi \cdot d\mathbf{S} = R_0^2 \chi^* \nabla_r \chi \, d\Omega$$

$$= R_0^5 \sum_{\substack{\lambda\mu \\ \lambda',\mu'}} \frac{\dot{\alpha}^*_{\lambda\mu}}{\lambda} \cdot \dot{\alpha}_{\lambda'\mu'} (r/R_0)^\lambda (r/R_0)^{\lambda'-1} Y^*_{\lambda\mu}(\theta, \varphi) Y_{\lambda'\mu'}(\theta, \varphi)$$

which upon integration over the unit sphere yields two Kronecker deltas $\delta_{\lambda,\lambda'}$ and $\delta_{\mu,\mu'}$, giving

$$T = \frac{R_0^5 \varrho_0}{2} \sum_{\lambda\mu} \frac{|\dot{\alpha}_{\lambda\mu}|^2}{\lambda}. \tag{6a}$$

Thus the expansion parameters $\alpha_{\lambda\mu}$ become the generalized coordinates in the sense of small-oscillation theory from which it is known that $T$ can be put into the general form [19]

$$T^L = \frac{1}{2} \sum_{\lambda\mu} B_{\lambda\mu} |\dot{\alpha}_{\lambda,\mu}|^2. \tag{6b}$$

Comparing Eqs. (6a) and (6b) gives the mass parameters

$$B_{\lambda\mu} = B_\lambda = \frac{\varrho_0 R_0^5}{\lambda}. \tag{7}$$

The kinetic energy is diagonal and the mass parameter is not a function of $\mu$ for oscillations about a spherical equilibrium shape. This would not be so if the oscillations were about a nonspherical equilibrium shape, say a spheroid†; in this case, $T$ will still be diagonal but the mass parameter will depend upon $|\mu|$ as well as $\lambda$. We shall return to this matter later.

In a similar manner we can show that for these generalized coordinates the potential energy is also in diagonal form for small displacements from a spherical shape

$$V = \frac{1}{2} \sum_{\lambda\mu} C_{\lambda\mu} \,|\, \alpha_{\lambda,\mu} \,|^2 , \tag{8}$$

where the coefficient $C_{\lambda\mu}$ is independent of $\mu$ and is in fact the sum of two terms, one due to the surface energy and the second due to the coulomb energy. In Appendix B this coefficient is shown to have the form [57]

$$C_\lambda = (\lambda - 1)(\lambda + 2)R_0{}^2 \mathscr{S} - \frac{3(\lambda - 1)(Ze)^2}{2\pi(2\lambda + 1)R_0} . \tag{9}$$

Since there is no viscosity term, the potential (8) is conservative. The foregoing comments about nonspherical distortions apply here as well.

The classical Lagrangian of the system is just the difference of Eq. (6b) and Eq. (8) from which a generalized momentum $\pi_{\lambda,\mu}$ conjugate to the coordinates $\alpha_{\lambda,\mu}$ can be defined in the usual manner, thus leading to a Hamiltonian for small oscillations about a spherical shape

$$H = \sum_\lambda H_\lambda = \frac{1}{2} \sum_{\lambda\mu} \left( \frac{1}{B_\lambda} \,|\, \pi_{\lambda\mu} \,|^2 + C_\lambda \,|\, \alpha_{\lambda,\mu} \,|^2 \right) , \tag{10a}$$

which is just the Hamiltonian for system of uncoupled harmonic oscillators with frequencies

$$\omega_\lambda{}^2 = C_\lambda / B_\lambda .$$

The angular momentum of the drop is just

$$\mathbf{L} = \int \varrho_0 \mathbf{r} \times \mathbf{v}(\mathbf{r}) \, d\tau . \tag{11a}$$

---

† Throughout this volume the term *spheroid* will be used exclusively for an ellipse of revolution. The term *ellipsoid* will be reserved for those surfaces whose cross sections parallel to the coordinate planes are all ellipses; that is, the three semiaxes are such that $a \neq b \neq c \neq a$.

By making use of the velocity potential (4) this expression can be written

$$\mathbf{L}^L = i \sum_{\lambda\mu\mu'} B_\lambda \dot{\alpha}^*_{\lambda\mu} \alpha_{\lambda\mu'} \langle \lambda\mu' \mid \mathbf{L} \mid \lambda\mu \rangle . \tag{11b}$$

Here the classical angular momentum matrix is defined as

$$i \langle \lambda'\mu' \mid \mathbf{L} \mid \lambda\mu \rangle \equiv \int Y^*_{\lambda'\mu'}(\theta, \varphi)(\mathbf{R}_0 \times \boldsymbol{\nabla}\chi)\, d\Omega ,$$

which is clearly diagonal in $\lambda$.

If the nuclear system possesses simple harmonic fluid oscillations, then we might suppose that the expression obtained by quantizing Eqs. (10) and (11b) would represent an adequate starting point from which to investigate low energy nuclear properties. This will be the case if the number of nucleons outside the deformable core is fairly small so that the energy of deformation of this core is smaller than the zero-point energy of the surface oscillations. That is, the nucleus will assume no permanent distortion and even-even nuclei (and the even-even cores of odd-$A$ and odd-odd nuclei) will be described by these quantized surface vibrations in a laboratory coordinate system [57, 91, 146].

In order to quantize the Hamiltonian of Eq. (10) we need only to demand that

$$[\pi_{\lambda,\mu} , \alpha_{\lambda',\mu'}] = -i\hbar\, \delta_{\lambda\lambda'}\, \delta_{\mu\mu'} , \tag{12a}$$

all other commutators being zero. It is customary to use the second quantized number representation [16] by introducing the usual creation and destruction operators $b^+_{\lambda,\mu}$ and $b_{\lambda,\mu}$, respectively. Here $b^+_{\lambda,\mu}$ operating on the vacuum state $\mid 0\rangle$ creates a $\lambda$ surfon in the state $\lambda\mu$

$$b^+_{\lambda\mu} \mid 0\rangle \quad = \mid \lambda, \mu\rangle$$

while

$$b_{\lambda\mu} \mid \lambda, \mu\rangle = \mid 0\rangle .$$

In terms of the generalized coordinates and momenta,

$$\begin{aligned} \alpha_{\lambda,\mu} &= \sqrt{\hbar/2B_\lambda\omega_\lambda}\; (b_{\lambda,\mu} + (-1)^\mu\, b^+_{\lambda-\mu}) \\ \pi_{\lambda,\mu} &= i\sqrt{\hbar B_\lambda \omega_\lambda/2}\; (b^+_{\lambda,\mu} - (-1)^\mu\, b_{\lambda-\mu}). \end{aligned} \tag{13}$$

Then, Eqs. (12a) and (13) imply that

$$[b^+_{\lambda,\mu}, b_{\lambda',\mu'}] = \delta_{\lambda\lambda'}\delta_{\mu\mu'} \qquad \text{(12b)}$$

the surfons, the quanta of the surface oscillations, are Bose particles. The number operator that counts the quanta in the state $|\lambda\mu\rangle$ is

$$n_{\lambda,\mu} = b^+_{\lambda\mu} b_{\lambda\mu}$$

while the operator that counts the total number of $\lambda$ surfons is

$$N_\lambda = \sum_{\mu=-\lambda}^{\lambda} n_{\lambda\mu} = 0,\ 1,\ 2,\ \ldots. \qquad \text{(10b)}$$

The Hamiltonian of Eq. (10) assumes the simple form

$$H = \sum_\lambda \hbar\omega_\lambda[N_\lambda + (2\lambda + 1)/2]. \qquad \text{(10c)}$$

The eigenvectors, in this notation are just $|\lambda, N, L, \mu\rangle$ and are found by successive applications of the creation operator $b^+_{\lambda\mu}$. Taking a normalizing constant $\mathcal{N}_{N,L}$ we find

$$|\lambda 1 L = \lambda, \mu\rangle = b^+_{\lambda\mu}|0\rangle$$

$$|\lambda 2 L\mu\rangle = \mathcal{N}_{2L} \sum_\nu C(\lambda\lambda L; \nu, \mu - \nu, \mu)\, b^+_{\lambda\mu-\nu} b^+_{\lambda\nu}|0\rangle$$

$$|\lambda 3 L'\mu\rangle = \mathcal{N}_{3L} \sum_{\nu\sigma} C(\lambda L L'; \sigma, \mu - \sigma, \mu) C(\lambda\lambda L; \nu, \sigma - \nu, \sigma) \times b^+_{\lambda\mu-\sigma} b^+_{\lambda\mu-\nu} b^+_{\lambda\nu}|0\rangle.$$

For the two-surfon case ($N = 2$), the normalizing constant is simply $\mathcal{N}_{2L} = 1/\sqrt{2}$ for all $\lambda$.

The angular momentum component $L_z$ also assumes a particularly simple form in this representation since

$$\langle\lambda\mu'|L_z|\lambda\mu\rangle = \mu\delta_{\mu\mu'}$$

and on using the definitions (13), Eq. (11b) yields

$$L_z = \hbar \sum_{\lambda\mu} \mu n_{\lambda\mu} = \sum_\lambda (L_z)_\lambda.$$

The $\lambda$ surfons are thus seen to be quanta with spin $\lambda$. It should be noted that this particularly simple form of the oscillator model is a direct result of the irrotational nature that is assumed for the fluid flow.

Now from the theory of nonrelativistic, quantized fields [9] it is known that field particles of angular momentum $\lambda$ have parity $(-1)^\lambda$. Therefore, to describe nucleon levels with positive parity it is necessary to use fields with $\lambda$ even, whereas, to describe nucleon levels with negative parity it is necessary to use fields with $\lambda$ odd. Also, since it is generally true that $\omega_{\lambda_i} < \omega_{\lambda_j}$ for $i < j$, the lowest levels in even-even nuclei should have positive parity whereas negative parity levels will be at higher energies. Also the lowest positive parity levels can be described by $\lambda = 2$ surfons and, of course, the lowest negative parity levels by $\lambda = 3$ surfons. This is what is observed in the so-called transitional or vibrational regions (that is, those regions of $A$ which are neither very close to closed shells nor in the regions of permanent deformation). Figure 2-1 shows a schematic energy level diagram for vibrational nuclei described by this model $\lambda \leq 4$; Table 2-1 gives the total angular momentum to which several $\lambda$ surfons can couple. Level structures such as these are found in the vibrational regions (for example, around cadmium), the ground and first

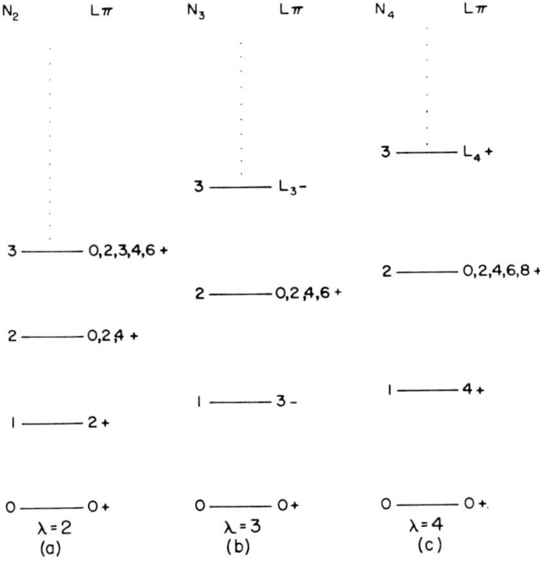


Fig. 2-1. A schematic diagram of the energy levels for the vibrational model without residual interactions for $\lambda \leq 4$. For $\lambda$ odd the parity alternates, being $(-1)^{N_\lambda}$. The possible values of $L_3$ and $L_4$ are given in Table 2-1. For even-even nuclei this model would predict that the low energy positive parity structure would be similar to the case (a), whereas the low energy negative parity structure would resemble case (b).

TABLE 2–1

The Total Angular Momentum $L$ to Which $N$ Surfons of Order $\lambda$ Can Couple

| $N$ | $\lambda = 2$<br>$L$ | $\lambda = 3$<br>$L$ | $\lambda = 4$<br>$L$ |
|---|---|---|---|
| 0 | 0 | 0 | 0 |
| 1 | 2 | 3 | 4 |
| 2 | 0, 2, 4 | 0, 2, 4, 6 | 0, 2, 4, 6, 8 |
| 3 | 0, 2, 3, 4, 6 | 1, 3, 3, 4, 5, 6, 7, 9 | 0, 2, 3, 4, 4, 5, 6, 6, 7, 8, 9, 10, 12 |
| . | . | . | . |
| . | . | . | . |
| . | . | . | . |

excited states of these nuclei being 0+ and 2+, respectively (as is true of almost all even-even nuclei). However, the degeneracy of the two-surfon state is removed by various residual particle interactions so that the second excited state is often observed as a 0, 2, 4+ triplet, the spins not necessarily in that order. Higher spin states of positive parity are frequently observed but the situation is not as clear cut. In many cases, near the first positive parity triplet a 3− state is to be found. Evidence for hexadecapole surfons is not seen in the low-lying energy level structures, but may be responsible for the enhanced $E4$ transitions that are observed from high energy electron scattering [113]. It is worthwhile pointing out that, in this degenerate model, the first $I = 3+$ level of the $\lambda = 2$ surfons appears well above the 0, 2, 4 triplet.

Nuclear models built upon theories of quantized surface oscillations have been extensively studied since their application to the problem of the deviations of nuclear magnetic dipole moments from the Schmidt limits [91]. In the following chapters we shall outline some of these models in order to delineate not only their shortcomings but to relate them to the more successful models applicable to the deformed regions.

It is of some interest to discuss the group $\mathscr{G}_\lambda$ under which the $H_\lambda$ of Eq. (10a) is invariant. The proper form with which to start is Eq. (10c) from which it is obvious that every one of the bilinear forms $b^+_{\lambda\mu'} b_{\lambda\mu''}$ commutes with $H_\lambda$; thus the generators of $\mathscr{G}_\lambda$ must be of this form. Also because these bilinear quantities preserve the symmetric bilinear forms $N_\lambda$ of Eq. (10b), $\mathscr{G}_\lambda$ is none other than $SU(2\lambda+1)$. Consider the set of operators

$$Q_{LM}^{(\lambda)} = A_\lambda^{(LM)} \sum_\mu C(\lambda L \lambda;\, \mu, M, \mu + M)\, b_{\lambda,\mu+M}^{+}\, b_{\lambda,\mu}\,;$$

and the coefficients $A_\lambda^{(LM)}$ need only be nonzero and finite (if in particular we pick $A_\lambda^{(1M)} = [\lambda(\lambda+1)]^{1/2}$, then the three $Q_{1M}^{(\lambda)}$ are the generators of $R(3)$). From the form of $Q_{LM}^{(\lambda)}$ we have $0 \leq L \leq 2\lambda$ and they number

$$n = \sum_{j=0}^{2\lambda} (2j+1) = (2\lambda+1)^2;$$

thus the $Q_{LM}^{(\lambda)}$ are the generators of $U(2\lambda+1)$. If we set $A_\lambda^{(00)} = 1$, then

$$Q_{00}^{(\lambda)} = \sum_\mu b_{\lambda\mu}^{+} b_{\lambda\mu} = \frac{H_\lambda}{\hbar\omega_\lambda} - \frac{2\lambda+1}{2}.$$

Therefore by removing this operator from the set we are left with the set containing $4\lambda(\lambda+1)$ operators, which are the generators of $SU(2\lambda+1)$.†

The regions of permanent nuclear deformation occur where both the neutron and proton shells are filling, that is, in regions where there are a considerable number of nucleons (or holes) outside of closed shells (for example, the rare earth and actinide regions). The nucleus stabilizes about some definite nonspherical nuclear shape with respect to which a fixed-axis system can be defined. The kinetic and potential energies can then be referred to this body-fixed system. A permanent deformation such as this will take place in those nuclear regions where the energy of deformation (change in coulomb and surface energies) is much larger than the zero-point vibrational energies. Indeed, if the energy of deformation is sufficiently large the system can, to first order, be considered a nonspherical rigid rotator, the effect of any vibrations being handled by perturbation methods.

By making use of the transformation (3) and the time derivatives of

---

† The author is indebted to Professor Derek L. Pursey and Dr. S. A. Williams, both of Iowa State University, for several discussions concerning these matters. The form of the operators $Q_{LM}^{(\lambda)}$ is due to Williams.

the $D^L_{MK}(\theta_i)$ which are shown in Appendix A to have the form

$$\dot{D}^L_{MK}(\theta_i) = i \sum_{k=1}^{3} \sum_{J=-L}^{L} \langle LJ \mid L_k \mid LK \rangle \, D^{L*}_{MJ}(\theta_i)\omega_k \,, \tag{14}$$

with $\omega_k$ the body-fixed components of the angular velocity, the kinetic energy equation (6b) becomes

$$T^B = \frac{1}{2} \sum_{\lambda} B_\lambda \Bigg\{ \sum_{\nu} \mid \dot{a}_{\lambda,\nu} \mid^2 + \sum_{\substack{\nu\nu' \\ kk'}} a^*_{\lambda\nu'} a_{\lambda\nu} \langle \lambda\nu' \mid L_{k'} L_k \mid \lambda\nu \rangle \, \omega_k \omega_{k'} + \Bigg[ i \sum_{\substack{\nu\nu' \\ k}} \dot{a}^*_{\lambda,\nu'} a_{\lambda,\nu} \langle \lambda\nu' \mid L_k \mid \lambda\nu \rangle \, \omega_k + \text{C.C.} \Bigg] \Bigg\}. \tag{15}$$

The three terms of the kinetic energy represent the vibrational energy, the rotational energy, and the rotation-vibration cross term, respectively. On defining the components of the inertial tensor $\mathscr{I}^{(\lambda)}_{k',k}$ by

$$\mathscr{I}^{(\lambda)}_{k',k} = B_\lambda \sum_{\nu',\nu} a^*_{\lambda\nu'} a_{\lambda\nu} \langle \lambda\nu' \mid L_{k'} L_k \mid \lambda\nu \rangle \,, \tag{16a}$$

the rotational kinetic energy can be placed in the more familiar form

$$T^B_{\text{rot}} = \frac{1}{2} \sum_{\substack{\lambda \\ k',k}} \mathscr{I}^{(\lambda)}_{k'k} \, \omega_{k'} \omega_k . \tag{17}$$

As yet, the body-fixed axis system has not been specified. In rigid-body dynamics we do this by picking that axis system which diagonalizes the inertial tensor. If the first term of Eq. (15) is not identically zero, this system will not be rigid; however, it has been shown that even for such nonrigid systems a sufficient condition to diagonalize the inertial tensor is to require $a_{\lambda,\nu}$ to be nonzero only for even or odd values of $\nu$ [160]. This condition also makes the rotation-vibration cross term of Eq. (15) identically zero, which is not unreasonable since we have arbitrarily reduced the number of degrees of freedom. (There is one exception: the case in which $\lambda = 2$ for which only two of the set of generalized coordinates $a_0$, $a_{\pm 1}$, $a_{\pm 2}$ can be independent and different from zero. It is easy to show from the ellipsoidal nature of the surface that the

$a_{\pm 1} \equiv 0$. Then, since $a_{-2} = a_2$, the five generalized coordinates are the three Euler angles, $a_0$, and $a_2$.)

The potential energy (8) when transformed into the body-fixed reference system becomes just

$$V^B = \frac{1}{2} \sum_{\lambda,\mu} C_\lambda \mid a_{\lambda\mu} \mid^2. \tag{18}$$

From earlier remarks concerning parity and the degree $\lambda$ of the surface expansion it should be clear that since parity is known to be a very good nuclear quantum number [130], the sums in Eqs. (15), (17), and (18) should be restricted to even $\lambda$ for positive parity states and to odd values of $\lambda$ for negative parity states.

On restricting the discussion to cases in which the rotation-vibration cross term can be neglected (there is only one detailed study in which this term has not been neglected and this study was restricted to $\lambda = 3$ surfaces [76]), we may define a $\lambda$th-order deformation parameter $\beta_\lambda$ by

$$a_{\lambda\mu} = \beta_\lambda \varepsilon_{\lambda\mu}, \tag{19}$$

where the $a_{\lambda\mu}$ have all been taken real and we may also require that

$$\sum_\mu \varepsilon_{\lambda\mu}^2 = 1. \tag{20}$$

The kinetic and potential energies then become

$$T^B = \frac{1}{2} \sum_\lambda B_\lambda(\dot{\beta}_\lambda^2 + \beta_\lambda^2 \sum_{\nu=-\lambda}^{\lambda} \dot{\varepsilon}_{\lambda\nu}^2) + \sum_{k=1}^{3} \mathscr{I}_k^{(\lambda)} \omega_k^2 \tag{21}$$

$$V^B = \frac{1}{2} \sum_\lambda C_\lambda \beta_\lambda^2 = \sum_\lambda V(\lambda) \tag{22}$$

and since these are referred to the principal axis system

$$\mathscr{I}_k^{(\lambda)} \equiv \mathscr{I}_{k,k}^{(\lambda)} = B_\lambda \beta_\lambda^2 \sum_{\nu\nu'} \varepsilon_{\lambda\nu'} \varepsilon_{\lambda\nu} \langle \lambda\nu' \mid L_k^2 \mid \lambda\nu \rangle = B_\lambda \beta_\lambda^2 i_k^{(\lambda)}. \tag{16b}$$

With the classical expressions for the kinetic and potential energies of this deformed drop we can proceed to quantize the system by well-known, but different, techniques from those used for the model of vibrating nuclei [35].

From the spherical harmonic expansion of the surface it is obvious that motion associated with different degrees $\lambda$ of that expansion will not interfere. Therefore, there is no loss of generality by restricting the ensuing discussion to a single value of $\lambda$.

The kinetic energy may be expressed in terms of the square of the time-rate change of the line element in a space with generalized coordinates $x_i$. If $G_{ij}$ is the covariant metric tensor, then the square of the line element is

$$(ds)^2 = \sum_{ij} G_{ij}\, dx^i\, dx^j ;$$

thus the kinetic energy is just

$$T = \frac{1}{2}\left(\frac{ds}{dt}\right)^2 = \frac{1}{2} \sum_{ij} G_{ij}\, \dot{x}^i\, \dot{x}^i. \tag{23}$$

By comparing Eqs. (21) and (23) the metric tensor is seen to have the form

$$G_{ij}(\lambda) = \begin{pmatrix} \mathbf{R}(\lambda) & \mathbf{0} \\ \mathbf{0} & \mathbf{D}(\lambda) \end{pmatrix}. \tag{24}$$

Here $\mathbf{D}(\lambda)$ is a diagonal matrix associated with the vibrational degrees of freedom

$$D_{ij}(\lambda) = B_\lambda \begin{pmatrix} 1 & & & & \mathbf{0} \\ & \beta_\lambda^2 & & & \\ & & \beta_\lambda^2 & & \\ & \mathbf{0} & & \ddots & \\ & & & & \beta_\lambda^2 \end{pmatrix} \tag{25}$$

and is $n \times n$ where

$$\begin{aligned} n &= (\lambda/2) + 1, \qquad \lambda \text{ even} \\ &= (\lambda + 1)/2, \qquad \lambda \text{ odd.} \end{aligned}$$

The matrix $\mathbf{R}(\lambda)$ is $3 \times 3$ and is associated with the kinetic energy of rotation. Since the angular velocity components $\omega_k$ are related to the time-rate change of the Euler angles $\dot{\theta}_i$ by

$$\omega_k = \sum_{=1}^{3} \sigma_{kj}(\theta_i)\dot{\theta}_j ,$$

the explicit form of the matrix elements $R_{ij}(\lambda)$ is

$$R_{ij}(\lambda) = B_\lambda \beta_\lambda^2 \sum_{k=1}^{3} \sigma_{ki}(\theta_i)\sigma_{kj}(\theta_i) i_k^{(\lambda)},$$

where Eq. (16b) has been used. The $R_{ij}(\lambda)$ are clearly symmetric.

The determinant of the metric tensor $G(\lambda)$ is just

$$G(\lambda) = B_\lambda^{n+3} \beta_\lambda^{2(n+2)} i_1^{(\lambda)} i_2^{(\lambda)} i_3^{(\lambda)} \sin^2 \theta_2 .$$

On recalling the definition of the contravariant metric tensor $G^{ij}(\lambda)$,

$$G^{ij}(\lambda) G_{jk}(\lambda) = \delta_{ik} ,$$

the Laplacian operator in this generalized coordinate system is

$$\nabla^2 = \sum_{l,m} \frac{1}{\sqrt{G}} \frac{\partial}{\partial x_l} \left( \sqrt{G}\, G^{lm} \frac{\partial}{\partial x_m} \right). \tag{26}$$

Since one form of the quantum condition is the replacement of the classical momentum $\mathbf{p}$ by $i \hbar \nabla$ and assuming suitable boundary conditions, the Hamiltonian for the system becomes

$$H_\lambda = -\frac{\hbar^2}{2} \sum_{l,m} \frac{1}{\sqrt{G(\lambda)}} \frac{\partial}{\partial x_l} \left( \sqrt{G(\lambda)}\, G^{lm}(\lambda) \frac{\partial}{\partial x_m} \right) + V(\lambda), \tag{27}$$

where the potential is that of Eq. (22).

In one case the Hamiltonian becomes quite simple and that is if the asymmetry parameters $\varepsilon_{\lambda,\mu}$ of Eq. (19) are taken as fixed so that the $\dot{\varepsilon}_{\lambda\mu} = 0$. Then the metric tensor (24) is just $4 \times 4$, since there is but one vibrational degree of freedom. We shall return to this discussion in the next chapter.

On the other hand, if all of the permissible vibrational degrees of freedom are available, then the metric tensor will not consist of only two diagonal blocks. Since the rotation-vibration term of Eq. (15) is not necessarily identically zero, there will be nonzero terms in the off-diagonal blocks. This is associated with the fact that only an instantaneous principal axis transformation can be carried out. These axes are not fixed in the body but migrate with time as a result of the asymmetry vibrations.

CHAPTER 3

# Models of Even-Even Nuclei

## A. General Considerations

Even-even nuclei, it is usually assumed, have the spins of like particles paired off and, at least for the low energy structure, none of the individual particles (or pairs) are excited. The Hamiltonian for such deformed nuclei can be taken to be that of Eq. (2-27); that is, the model is that of a rotating and vibrating liquid drop. The solution of the associated Schrödinger equation will yield energy eigenvalues and state functions. From the latter, other physical properties can be calculated, all of which can be compared with experiment to test the validity of the model.

From the form of the metric tensor of Eq. (2-24) it is clear that the Schrödinger equation will separate into rotational and vibrational parts. In what follows, first the rotational solutions will be discussed and then the vibrational solutions will be developed.

We may solve the rotational problem by using the quantization method discussed in the last chapter and obtaining, from Eq. (2-26), the Laplacian operator for the rotational degrees of freedom [35]. Alternately, we may use an angular momentum representation, defining the total nuclear angular momentum **L** for these even-even nuclei (for odd-$A$ and odd-odd nuclei **L** will be the angular momentum of the core) having the usual commutation relations

$$[L_x, L_y] = iL_z, \text{ cyclically,} \tag{1a}$$

from which it follows that the body-fixed components satisfy the commutation relations (Appendix A)

$$[L_1, L_2] = -iL_3, \text{ cyclically.} \tag{1b}$$

The state vectors in this angular momentum representation will be denoted by $| LMK\rangle$ and will be diagonal in $L_z$ and $L_3$, the projections of $\mathbf{L}$ on the laboratory and body-fixed $Z$ axes, as well as in $\mathbf{L}^2$

$$\mathbf{L}^2 \,|\, LMK\rangle = L(L+1) \,|\, LMK\rangle \tag{2a}$$

$$L_z \,|\, LMK\rangle = M \,|\, LMK\rangle \tag{2b}$$

$$L_3 \,|\, LMK\rangle = K \,|\, LMK\rangle. \tag{2c}$$

These relations are shown schematically in Fig. 3-1.

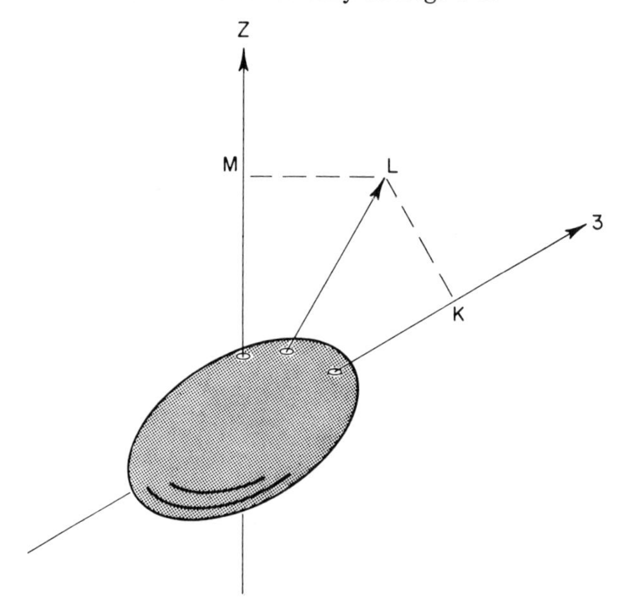


Fig. 3-1. The relation between the total angular momentum for an even-even nucleus $\mathbf{L}$, its projection $M$ on the laboratory $Z$ axis, and its projection $K$ on the body-fixed 3 axis. For symmetric systems this latter is taken along the axis of symmetry.

On taking the body-fixed axis system as the principal axis system, the Hamiltonian for a deformed, rigid rotator is

$$H = \frac{\hbar^2}{2}\left[\frac{L_1{}^2}{\mathscr{I}_1} + \frac{L_2{}^2}{\mathscr{I}_2} + \frac{L_3{}^2}{\mathscr{I}_3}\right], \tag{3}$$

again where the $\mathscr{I}_i$ are the principal moments of inertia.

One physically important case whose solution can be given in closed form is that of the axially symmetric rotator. On taking the 3 axis as the symmetry axis, then such symmetry implies

$$\mathscr{I}_1 = \mathscr{I}_2 = \mathscr{I}_0 \neq \mathscr{I}_3.$$

The Schrödinger equation with the Hamiltonian (3) is

$$H(\mathscr{I}_1 = \mathscr{I}_2) \mid LMK\rangle = E_{\text{sym}} \mid LMK\rangle$$

with

$$E_{\text{sym}} = \frac{\hbar^2}{2}\left[\frac{L(L+1)}{\mathscr{I}_0} + \left(\frac{1}{\mathscr{I}_3} - \frac{1}{\mathscr{I}_0}\right) K^2\right]; \tag{4}$$

thus the state functions $\mid LMK\rangle$ of (2) are the eigenfunctions of the symmetric top problem. (They are in fact the functions $D^{I*}_{MK}(\theta_i)$. For a general discussion see Rose [15].)

Since the general rotator possesses no axis of symmetry, its state functions, denoted by $\mid LM\rangle$, will be of the form

$$\mid LM\rangle = \sum_{K=-L}^{L} A_K \mid LMK\rangle. \tag{5}$$

By making use of the pseudospherical coordinate system (Appendix A), the nonvanishing matrix elements of **L** in the body-fixed system are

$$\langle LMK' \mid L_\mu \mid LMK\rangle = \sqrt{L(L+1)}\, C(L1L;\, K-\mu,\, \mu,\, K)\, \delta_{K',K-\mu}. \tag{6}$$

Therefore, the Hamiltonian (3) for such a rotator will only connect states in which $\Delta K = 0, \pm 2$. Thus the state functions for the most general rigid top can be classified into two broad categories: one for which the $K$-quantum numbers of Eq. (5) are all even, and the other for which they are all odd.

A further classification of these state functions arises from the fact that the assignment of the body-fixed coordinate axis labels 1, 2, 3 is arbitrary so that no physical quantity will be changed upon relabeling the semiaxes of the momental ellipsoid (that is, by picking a different principal axis system from among the six semiaxes of this ellipsoid). There are 24 ways in which we can make this choice, providing we use only right-handed systems (there are also 24 left-handed systems). Therefore, we can transform from one choice to another without the physical content of the problem being changed. In order to investigate the symmetries that arise, we define the relabeling transformations $T_1$, $T_2$, and $T_3$ [56]:

(i) $T_1$ produces the interchange $1 \leftrightarrow -1$, $3 \leftrightarrow -3$. It is equivalent to a rotation of $\pi$ about the 2 axis. $T_1^2 = 1$.
(ii) $T_2$ produces the interchange $1 \leftrightarrow 2$, $2 \leftrightarrow -1$. It is equivalent to a rotation of $\pi/2$ about the 3 axis. $T_2^4 = 1$.
(iii) $T_3$ produces the interchange $1 \leftrightarrow 2 \leftrightarrow 3 \leftrightarrow 1$. $T_3^3 = 1$.

The most general relabeling transformation that does not involve a change of handedness is $T_1^i T_2^j T_3^k$. Here $i = 1, 2$; $j = 1, 2, 3, 4$; $k = 1, 2, 3$; thus we get the required 24 different principal axis systems.

Applying $T_1$ and $T_2^2$ to the state function (5) yields

$$T_1 \mid LM\rangle = \pm \mid LM\rangle$$
$$T_2^2 \mid LM\rangle = \pm \mid LM. \tag{7}$$

There are then four classes of rigid rotator state functions labeled by the sign combinations of the transformations (7). These classes are the four representations $A$, $B_1$, $B_2$, $B_3$ of the point group $\mathbf{D}_2$ to which the system that is defined by Eq. (1) and (3) belongs [11].

In Appendix A we show that for the symmetric-top state functions

$$T_1 \mid LMK\rangle = e^{i\pi(L+K)} \mid LM-K\rangle$$
$$T_2^2 \mid LMK\rangle = e^{i\pi K} \mid LMK\rangle. \tag{8}$$

Using these relations and the preceding ones, we can relate the expansion coefficients of Eq. (5) for negative $K$ to those with positive $K$:

*K even*

(i) $A$-representation

$A_{-k} = (-1)^L A_K$

(ii) $B_1$-representation

$A_{-K} = -(-1)^L A_K$.

*K odd*

(iii) $B_2$-representation

$$A_{-K} = -(-1)^L A_K$$

(iv) $B_3$-representation

$$A_{-K} = (-1)^L A_K .$$

From these relations it follows that the energy level structure of $A$-type systems will contain one level with $L = 0$, no levels with $L = 1$, two levels with $L = 2$, one with $L = 3$, three with $L = 4$, etc. The $B$-type systems will have no levels with $L = 0$, one each with $L = 1$ and 2, two each with $L = 3$ and 4, three each with $L = 5$ and 6, etc. In Table 3-1, all of these relations that are associated with the four different symmetry types are tabulated. It should be emphasized that these spin assignments are quite general and are completely independent of the particular form or model we choose for the momental ellipsoid.

TABLE 3–1

Symmetry Properties of the State Functions of the General, Deformed Rigid Rotator; the Representation of the Point Group $\mathbf{D}_2$ to Which They Belong; and the Number of Allowed Values of the Angular Momentum. These Properties Are Independent of the Model Chosen for the Momental Ellipsoid

| Representation | $K$ | $A_K/A_{-K}$ | $T_1 \mid LM\rangle$ | $T_2^2 \mid LM\rangle$ | Number of allowed $L$ values: $L$ even | $L$ odd |
|---|---|---|---|---|---|---|
| $A$ | even | $(-1)^L$ | + | + | $(L+2)/2$ | $(L-1)/2$ |
| $B_1$ | even | $-(-1)^L$ | − | + | $L/2$ | $(L+1)/2$ |
| $B_2$ | odd | $-(-1)^L$ | + | − | $L/2$ | $(L+1)/2$ |
| $B_3$ | odd | $(-1)^L$ | − | − | $L/2$ | $(L+1)/2$ |

In fact, a great deal more can be done without detailing the model of the momental ellipsoid. The state with the smallest eigenvalue in the $A$ representation is that for which $L = 0$

$$\mid 00\rangle = \mid 000\rangle ,$$

the energy eigenvalue being identically zero for all moments of inertia.

The states with the next higher angular momentum are the two with $L = 2$. On calling $N$ the ordinal of the angular momentum state, then the state functions for these $L = 2$ states have the form

$$| 2MN\rangle = A_0^N \,| 2M0\rangle + (A_2^N/\sqrt{2})\,(| 2M2\rangle + | 2M-2\rangle)$$

with energy eigenvalues

$$\begin{aligned} E(2N, A) = \hbar^2 \left(\frac{1}{\mathscr{I}_1} + \frac{1}{\mathscr{I}_2} + \frac{1}{\mathscr{I}_3}\right) & \\ + (-1)^N \hbar^2 \Bigg[\left(\frac{1}{\mathscr{I}_1}\, \frac{1}{\mathscr{I}_2}\, \frac{1}{\mathscr{I}_3}\right)^2 & \\ - 3\left(\frac{1}{\mathscr{I}_1\mathscr{I}_2} + \frac{1}{\mathscr{I}_1\mathscr{I}_3} + \frac{1}{\mathscr{I}_2\mathscr{I}_3}\right)\Bigg]^{1/2}, & \end{aligned} \tag{9}$$

the ordinal $N$ being assigned from the lowest to the highest (that is, if $N < N'$ then $E(IN) < E(IN')$). The $L = 3$ state has a single-component state function

$$| 3M\rangle = (1/\sqrt{2})\,(| 3N2\rangle - | 3M-2\rangle)$$

and eigenvaluc

$$E(3, A) = 2\hbar^2 \left(\frac{1}{\mathscr{I}_1} + \frac{1}{\mathscr{I}_2} + \frac{1}{\mathscr{I}_3}\right). \tag{10}$$

Equations (9) and (10) are invariant under a cyclic interchange of the moments of inertia. This is generally true for $A$-type systems. Also from these two equations a model independent relation for these $A$-type systems is easily obtained

$$E(21, A) + E(22, A) = E(3, A)\,. \tag{11a}$$

This relation is often useful to see whether or not an experimental spectrum is possibly that of an asymmetric rotator.

We can proceed in this manner for higher values of $L$. However, only for the case $L = 5$ is the secular equation a quadratic and the eigenvalues easily expressed in closed form. For the other cases, the eigenvalues and state vector components (the $A_K$) are most easily obtained by a machine diagonalization of the Hamiltonian (3) in the basis (2).

Although it is not always possible to obtain explicit expressions for the eigenvalues, we can derive general relations [161] similar to (11a). The diagonal matrix elements of the $L_i^2$ in the representation (2) are just

$$\langle LMK \mid L_{1,2}^2 \mid LMK \rangle = \tfrac{1}{2}[L(L+1) - K^2]$$
$$\langle LMK \mid L_3^2 \mid LMK \rangle = K^2.$$

We now make use of the fact that the trace of a matrix is equal to the sum of its eigenvalues

$$\sum_N E(L, N) = Tr \langle LMK \mid H \mid LMK' \rangle$$

$$= \frac{\hbar^2}{4}\left[\left(\frac{1}{\mathcal{I}_1} + \frac{1}{\mathcal{I}_2}\right)L(L+1)\sum_{K=0}^{L} 1 + \left(\frac{2}{\mathcal{I}_3} - \frac{1}{\mathcal{I}_1} - \frac{1}{\mathcal{I}_2}\right)\sum_{K=0}^{L} K^2\right].$$

For $L$ even

$$N = \frac{L+2}{2}$$

$$\sum_{K=0}^{L} K^2 = \frac{L(L+1)(L+2)}{6}$$

whereas for $L$ odd

$$N = \frac{L-1}{2}$$

$$\sum_{K=0}^{L} K^2 = \frac{(L-1)\, L(L+1)}{6}$$

since for $A$-type systems $K$ is even. Therefore, for $L$ even

$$\sum_N E(L, N) = \frac{\hbar^2}{12} L(L+1)(L+2) \left(\frac{1}{\mathcal{I}_1} + \frac{1}{\mathcal{I}_2} + \frac{1}{\mathcal{I}_3}\right)$$

whereas for $L$ odd

$$\sum_{N'} E(L, N') = \frac{\hbar^2}{12} (L-1)L(L+1) \left(\frac{1}{\mathcal{I}_1} + \frac{1}{\mathcal{I}_2} + \frac{1}{\mathcal{I}_3}\right),$$

from which it is obvious that

$$\sum_N E(L, N) = \sum_{N'} E(L+1, N'), \qquad L \text{ even}, \tag{11b}$$

a generalization of Eq. (11a). Other relations can be derived in a similar manner. For example, using the relation

$$\sum_{\text{pairs}N,N'} E(L, N)\,E(L, N') = \frac{1}{2}\{[Tr\,\langle LMK \mid H \mid LMK'\rangle]^2 - Tr\,\langle LMK \mid H^2 \mid LMK'\rangle\}$$

we can show that [161]

$$\begin{aligned} E(5,1) &= 4E(2,1) + E(2,2) \\ E(5,2) &= 4E(2,2) + E(2,1). \end{aligned} \tag{11c}$$

As will be seen later, these relations become inequalities when vibrations are added to the system.

The lowest spin for the $B$-type systems is $L = 1$ and the normalized state functions are

$$\mid 1M\rangle = \mid 1M0\rangle, \qquad B_1\text{-representation} \tag{12a}$$

$$\mid 1M\rangle = (1/\sqrt{2})\,(\mid 1M1\rangle + \mid 1M-1\rangle, \qquad B_2\text{-representation} \tag{12b}$$

$$\mid 1M\rangle = (1/\sqrt{2})\,(\mid 1M1\rangle - \mid 1M-1), \qquad B_3\text{-representation} \tag{12c}$$

the energy eigenvalues being

$$E(1, B_1) = \frac{\hbar^2}{2}\left(\frac{1}{\mathscr{I}_1} + \frac{1}{\mathscr{I}_2}\right)$$

$$E(1, B_2) = \frac{\hbar^2}{2}\left(\frac{1}{\mathscr{I}_1} + \frac{1}{\mathscr{I}_3}\right)$$

$$E(1, B_3) = \frac{\hbar^2}{2}\left(\frac{1}{\mathscr{I}_2} + \frac{1}{\mathscr{I}_3}\right).$$

The eigenvalues belonging to the $B_2$ and $B_3$-representations can be gotten from that for the $B_1$-representation by replacing $\mathscr{I}_2$ by $\mathscr{I}_3$ and $\mathscr{I}_1$ by $\mathscr{I}_3$, respectively. This is also true for the other values of $L$.

The three $L = 2$ state functions contain but one component. The state function associated with the $B_1$-representation contains only a term with $\mid K \mid = 2$ whereas the other two state functions contain only terms with $\mid K \mid = 1$. For the $B_1$-representation

$$\mid 2M\rangle = (1/\sqrt{2})\,(\mid 2M2\rangle - \mid 2M-2\rangle),$$

the energy eigenvalue being

$$E(2, B_1) = \frac{\hbar^2}{2}\left(\frac{4}{\mathscr{I}_3} + \frac{1}{\mathscr{I}_1} + \frac{1}{\mathscr{I}_2}\right).$$

The other two eigenvalues can be obtained by the proper interchange of $\mathscr{I}_1$, $\mathscr{I}_2$, and $\mathscr{I}_3$. Other eigenvalues can be obtained similarly; but for $I > 4$, the secular equations are again not quadratic and the eigenvalues and components of the eigenvectors are more easily obtained numerically. The energy eigenvalues for the general rigid rotator have been calculated in terms of the moment of inertia ratios and the results tabulated [18, 115].

Nothing as yet has been said about the relation between symmetry type and the parity of the levels. The parity of nuclear states is taken relative to the parity of the ground states of even-even nuclei, which is assigned positive parity. Since the spins of such states are zero, only rotational models using the $A$-representation are permissible. The problem of assigning a symmetry type to the negative parity level systems in even-even nuclei then arises since many deformed nuclei do possess negative parity bands with the rotational spacing. It is possible to relate parity and representation through the following argument [77]. Consider the operator $C$ of complex conjugation. In the context of angular momentum theory, $C$ is analogous to reflection in the plane $y = 0$. On using the angular momentum stepping operators (defined in the laboratory system)

$$CL_\pm = C(L_x \pm iL_y) = L_x \mp iL_y = L_\mp ,$$

which is equivalent to the reflection $y \to -y$. Now the parity transformation may be defined either as the reflection in the origin or equivalently as a reflection in any of the coordinate planes times a rotation of $\pi$ about the coordinate axis normal to the plane of reflection. The parity operator $P$ is then just

$$P = R_y C = e^{i\pi L_y} C.$$

Using the properties of the $D^L_{MK}$ (Appendix A), we have

$$\begin{aligned} PD^{L*}_{MK}(\theta_i) &= R_y C D^{L*}_{MK}(\theta_i) = (-1)^{L+M} D^L_{-MK}(\theta_i) \\ &= (-1)^{L-K} D^{L*}_{M-K}(\theta_i). \end{aligned}$$

Using this relation, the form of the rotational state functions of Eq. (5) and the relation between $A_K$ and $A_{-K}$ from Table 3-1 for each representation, we find that

$$P \mid LM\rangle = + \mid LM\rangle, \qquad A \text{ and } B_2 \text{ symmetries}$$
$$P \mid LM\rangle = - \mid LM\rangle, \qquad B_1 \text{ and } B_3 \text{ symmetries.}$$

This first relation is consistent with the observation of $0+$ ground states in even-even nuclei. The second relation restricts models of negative parity states in these nuclei to either a $K$-even or a $K$-odd representation. Experimental evidence indicates that these negative parity states are predominately of even $K$ so that model studies should be restricted to the $B_1$-representation. Published studies of negative parity models have invariably made this choice.

In order to complete this model-independent discussion it is necessary to consider what happens to the energy eigenvalues as the rigidity condition is relaxed. In analogy with classical systems we might expect that if the system is not perfectly rigid the equatorial diameter will increase slightly as it rotates, and as a consequence of the constant volume condition the polar diameter will decrease. The effect on the moments of inertia would be similar; thus the energy levels would each be lowered slightly. For symmetric tops in their $K = 0$ bands the increase in the equatorial diameter increases $\mathscr{I}_0$; therefore expressing the eigenvalue in a power series in $L(L+1)$ we would have instead of Eq. (4)

$$E_{\text{sym}}(L, K=0) = (\hbar^2/\mathscr{I}_0)L(L+1) - bL^2(L+1)^2 + cL^3(L+1)^3 + \cdots \tag{13a}$$

Here $b$ is a small, positive quantity and all of these expansion coefficients are model dependent. (The three-parameter expansion (13a) has been used in the study of the energy level structures of some deformed even-even nuclei [150].) A relation similar to (13a) is valid for asymmetric systems and, in fact, can be written as [127]

$$E_v(L, N) = E(L, N) - bE^2(L, N) + \cdots, \tag{13b}$$

where the subscript $v$ denotes the eigenvalue including vibration. For small deviations from rigidity, $b$ can be shown to be positive, small, and independent of $L$. Relaxing the rigidity condition depresses each energy level; the higher the level the greater the depression. Thus relation (11a) becomes

$$E(21,A) + E(22,A) \geq E(3,A). \tag{11d}$$

It has been suggested by Mallman [128] that all even-even nuclei show evidence of being almost rigid rotators. We should expect that the lowest rotational bands, if they existed at all in such nuclei, would belong to the; $A$-representation with positive parity since these states have the maximum possible symmetry, which is a requirement of the lowest energy eigenstates. Mallman tacitly made this assumption and used the expressions (11a) and (11d) to test the rotational model. A more recent investigation [28] used the experimental data from 36 nuclei in the range $24 \leq A \leq 254$, Eq. (11a) in the form

$$\frac{E(21) + E(22) - E(31)}{E(21) + E(22)}, \tag{14}$$

and Eq. (11c) in a similar form. The spectra can be considered rotational if this ratio is of the order of 1 or 2% or less. With this criterion the rotational regions are quite sharply delineated: one is in the rare earth region terminating in the platinum isotopes ($150 \leq A \leq 194$); the other is generally in the actinide region $A \geq 226$. Also $Mg^{24}$ and $Ba^{134}$ can be considered, on this basis, as being deformed rotators. One interesting feature of this investigation is that near the edges of the lower A deformed region, the ratio (14) is small but negative, which is a violation of the model condition (11d). The deformed regions, as they are called, can also be identified from the fact that the first excited states ($I = 2$) are much lower for nuclei inside these regions than for nuclei outside them. This can be seen in Fig. 3-2 which gives the energies of these states. It can be seen that the onset of nuclear deformation is quite sudden in both regions.

Although agreement with relations (11a), (11c), or others similar, is evidence for rotational motion, it is useful—if the relations are satisfied—to determine the moment of inertia ratios from the lower levels and then use these to check higher angular momentum states. To be consistent it is necessary to take into account vibrations as are seen in (13b) so that the fitting procedure uses four parameters (two moment of inertia ratios, a vibrational parameter $b$, and a scale factor). Mallman [128] has done this (it is also necessary to investigate the electromagnetic

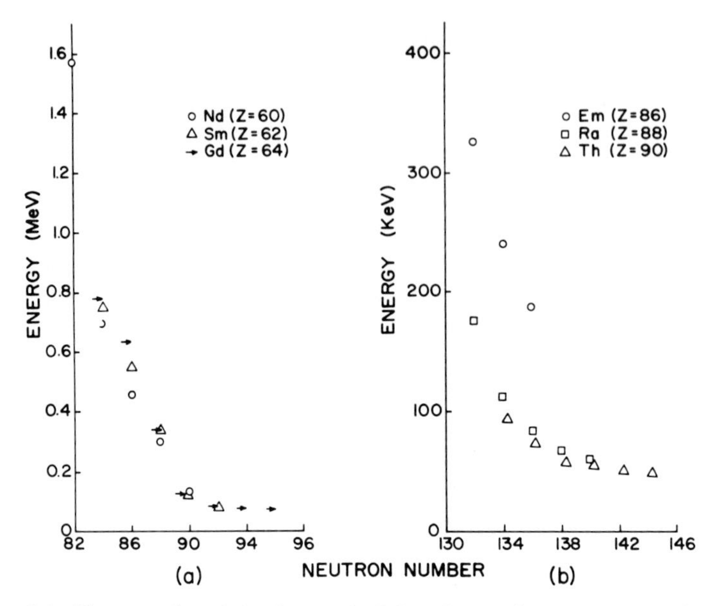


Fig. 3-2. The energies of the first excited $L = 2$ state in even-even nuclei, plotted against neutron number $N$ near the lower edge of the deformed regions. In the rare earth region rotational spectra are first observed at $N = 90$ whereas in the actinides they appear at $N = 138$.

properties of these states, which he did; see Chapter 6) and finds the further significant fact that in those regions which are rotational, the moment of inertia ratios are quite close to the hydrodynamic values of Eq. (2-16) with $\lambda = 2$.

## B. Models with Axial Symmetry

Two alternate collective models of deformed even-even nuclei have been proposed that are perhaps more appropriately called models of the nuclear momental ellipsoid since many of the predictions that are common to these models arise from symmetry considerations. The most simple and certainly the most useful model from an experimental point of view is the one developed by Bohr and Mottelson [56, 57] that uses the hydrodynamic moments of inertia and restricts the surface to $\lambda = 2$. There are then five degrees of freedom and for the two vibrational degrees it is customary to define the $\varepsilon_{\lambda\mu}$ of Eq. (2-19) as

$$\begin{aligned} \varepsilon_{2,0} &= \cos\gamma \\ \varepsilon_{2,\pm 2} &= (1/\sqrt{2}\,)\sin\gamma . \end{aligned} \tag{15}$$

With this choice the moments of inertia, (2-16b) become

$$\mathscr{I}_k^{(2)} = 4B_2\beta_2{}^2\sin^2(\gamma - 2\pi k/3). \tag{16}$$

A further assumption of this model is that these nuclei are, on the average, symmetric: that is, $\bar{\gamma} = 0$, from which it follows that

$$\begin{aligned} &\mathscr{I}_1^{(2)} = \mathscr{I}_2^{(2)} = \mathscr{I}_0^{(2)} = 3B_2\beta_2{}^2 \\ &\mathscr{I}_3^{(2)} = 0 . \end{aligned} \tag{17}$$

Therefore, Eq. (4) will be meaningful only if $K$ is taken identically zero. This yields what is often known as the ground state rotational band, the energy eigenvalues being

$$E(L) = \frac{\hbar^2}{6B_2\beta_2{}^2} L(L+1), \qquad K = 0. \tag{18}$$

Because of the $A$-type representation requirements, $L$ must be taken even. This is in quite good agreement with the observed sequence in even-even nuclei, which is $0+$, $2+$, $4+$, $6+$, .... Levels up to $L = 18$ have been observed by various reactions and thus the $L(L+1)$ rule is quite well satisfied when allowance is made for the deformation vibrations.

This model will have two higher, or excited, bands (it will, of course, have an infinite number of excited bands but only the two mentioned subsequently have been positively identified and extensively used). The $\beta$ band is associated with vibrations in the $\beta_2$ degree of freedom and because these are not associated with intrinsic rotation it will be a $K = 0$ band, the level sequence being given by Eq. (18) and the band head by $\hbar\,\omega_\beta$. However, the value of $\beta_2$ will be greater in this band than in the ground state band. The second excited band is often called the "gamma band" and is associated with oscillations in the $\gamma$ degree of freedom as it is defined in (15). For a vibration such as this $\bar{\gamma}$ will not be identically zero. Therefore Eq. (4) can be used and the lowest levels will be associated with $K = \pm 2$; thus the spin sequence is $L = 2+$, $3+$, $4+$, $5+$, .... If more than two levels of a given spin appear (for $L > 2$) which are rotational, higher bands based on other values of $K$ are necessary. As many as four, or more, $K$ bands have been assigned to classify the experimental data in some deformed even-even nuclei. Figure 3-3 is a schematic drawing of how these bands would appear in this type of nucleus. The negative parity bands in this model will be discussed later.

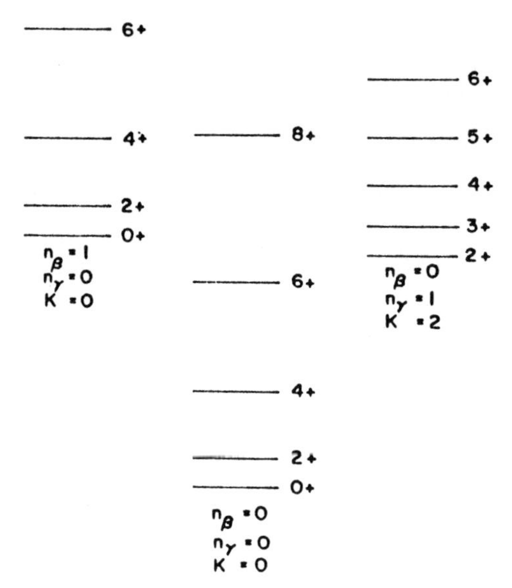


Fig. 3-3. An energy level diagram for a symmetrically deformed nucleus showing the $K = 0$ bands (the ground and beta bands) as well as the $K = 2$ or gamma band.

It is quite easy to obtain an analytical expression for the vibration parameter $b$ of Eq. (13a) in a simple, semiclassical way provided that the energy of rotation is much less than the energy of vibration [33]. If the nucleus can vibrate and rotate, then the energy may be written

$$E(L, \beta) = \frac{\hbar^2}{2B_2\beta_2^2} L(L+1) + \frac{1}{2} C_2(\beta_2 - \beta_{20})^2. \tag{19}$$

If the nucleus were not rotating, $\beta_{20}$ would be its equilibrium deformation; however, rotation induces a slight elongation of the equatorial diameter (centrifugal stretching), which changes this deformation from $\beta_{20}$ to $\bar{\beta}$. The new equilibrium deformation can be found by setting

$$\partial E(L, \beta_2)/\partial\beta_2 = 0$$

and solving for $\bar{\beta} = \beta_{20} + \Delta$, or for $\Delta$:

$$\Delta = \frac{\hbar^2 L(L+1)\beta_{20}}{3[\hbar^2 L(L+1) + C_2 B_2 \beta_{20}^4]},$$

which is substituted into (19). The parameter $b$ is then just the coefficient of the $L^2(L+1)^2$ term

$$b = -\frac{12}{(\hbar\omega_\beta)^2}\left(\frac{\hbar^2}{2\mathscr{J}_0^{(2)}}\right)^3. \tag{20}$$

Gamma vibrations will also make a contribution to $b$; however, in order to find the exact influence of one set of bands on another, a more careful treatment is necessary and illuminating.

Although the ensuing analytical discussion of the symmetric-core model follows the treatment of Birbrair et al. [51] and Lipas [123], it has been extended to studies of negative parity states in these nuclei. This analysis utilizes the hydrodynamic moments of inertia with the parameters $\beta$ and $\gamma$. Another approach, not specifying the $\varepsilon_{2\mu}$ of Eq. (2-19), but utilizing only the $a_{20}$, $a_{2\pm 2}$ parameters may also be used [87]. To the same degree of approximation the results, of course, are identical. It is assumed that the $\beta$ and $\gamma$ oscillations are small in amplitude and high in frequency. Therefore, the equilibrium shape is characterized by the mean value of $\beta_2 = \beta_{20}$ and $\gamma = 0°$, the moments of inertia being

$$\begin{gathered}\mathscr{J}_1^{(2)} \cong \mathscr{J}_2^{(2)} \cong 3B_2\beta_{20}^2 = \mathscr{J}_0^{(2)} \\ \mathscr{J}_3^{(2)} \cong 4B_2\beta_{20}^2\gamma^2.\end{gathered} \tag{21}$$

In lowest order the vibrations and rotations are almost completely uncoupled (they would be completely so except for the peculiar nature of $\mathscr{J}_3^{(2)}$ that vanishes as $\gamma^2$); thus this order is often called the "adiabatic approximation."

By using the techniques outlined in the last chapter, the vibrational kinetic energy is

$$T_{\text{vib}} = -\frac{\hbar^2}{2}\left[\frac{1}{B_2}\frac{\partial^2}{\partial\beta_2{}^2} + \frac{1}{B_\gamma}\frac{1}{\gamma}\frac{\partial}{\partial\gamma}\left(\gamma\frac{\partial}{\partial\gamma}\right)\right], \tag{22a}$$

where $B_\gamma = B_2\beta_{20}^2$. The rotational kinetic energy is

$$T_{\text{rot}} = \frac{\hbar^2}{2\mathscr{J}_0^{(2)}}[L(L+1) - K^2] + \frac{\hbar^2}{2}\frac{K^2}{\mathscr{J}_3^{(2)}(\gamma)}. \tag{23a}$$

The Hamiltonian is formed from the sum of these two terms and the vibrational potential energy which, for small vibrations about the assumed equilibrium shape, is

$$V(\beta, \gamma) = (1/2)\, C_\beta(\beta_2 - \beta_{20})^2 + (1/2)C_\gamma\gamma^2. \tag{24}$$

The coefficients $C_\beta$ and $C_\gamma$ are usually considered fitting parameters since they are related to the various band-head energies. However, an analytic expression for them can be obtained by making a Taylor series expansion of the potential function about the equilibrium shape. The coefficients of the quadratic terms are the required parameters.

The Schrödinger equation is most easily dealt with if the last term in Eq. (23a) is either included in the rotation-vibration interaction term [87] (which also includes the effect of the difference between $\mathscr{J}_1^{(2)}$ and $\mathscr{J}_2^{(2)}$) or placed in the vibrational kinetic energy term by redefining the rotational and vibrational kinetic energies to be

$$T'_{\text{rot}} = T_{\text{rot}} - \frac{\hbar^2}{2}\frac{K^2}{\mathscr{J}_3^{(2)}(\gamma)} \tag{22b}$$

$$T'_{\text{vib}} = T_{\text{vib}} + \frac{\hbar^2}{2}\frac{K^2}{\mathscr{J}_3^{(2)}(\gamma)}. \tag{23b}$$

The solutions to the Schrödinger equation

$$H\Psi(\theta_i, \beta, \gamma) = E\Psi(\theta_i, \beta, \gamma) \tag{25}$$

have the form

$$\Psi(\theta_i, \beta, \gamma) = f(\beta)\, g(\gamma) \mid LMK\rangle. \tag{26}$$

By substituting (26) into (25) it is seen that $f(\beta)$ and $g(\gamma)$ must satisfy

$$\left[-\frac{\hbar^2}{2B_2}\frac{d^2}{d\beta^2} + \frac{1}{2}C_\beta(\beta - \beta_0)^2\right] f(\beta) = E_\beta f(\beta) \tag{27}$$

$$\left\{-\frac{\hbar^2}{2B_\gamma}\left[\frac{1}{\gamma}\frac{d}{d\gamma}\left(\gamma\frac{d}{d\gamma}\right) - \left(\frac{K}{2\gamma}\right)^2\right] + \frac{1}{2}C_\gamma\gamma^2\right\} g(\gamma) = E_\gamma g(\gamma). \tag{28}$$

Equation (27) is the Schrödinger equation for a one-dimensional harmonic oscillator of mass $B_2$ oscillating about the equilibrium position $\beta_0$. On defining $\bar{\beta} = \beta - \beta_0$ its solutions are

$$f(\beta) = N_\beta H_{n_\beta}(\sqrt{B_2 \omega_\beta / \hbar}\, \bar{\beta}) \exp(-B_2 \omega_\beta \bar{\beta}^2 / 2\hbar) \tag{29a}$$

$$E_\beta = \hbar\, \omega_\beta (n_\beta + \tfrac{1}{2}), \qquad n_\beta = 0, 1, 2, \ldots, \tag{29b}$$

where the frequency is $\omega_\beta{}^2 = C_\beta / B_2$; $H_{n_\beta}(x)$ is a Hermite polynomial of order $n_\beta$; and the normalization constant $N_\beta$ is fixed by the requirement,

$$\int_{-\infty}^{\infty} f^2(\beta)\, d\bar{\beta} = 1. \tag{29c}$$

Equation (28) can be put into the form of the radial equation of the two-dimensional harmonic oscillator by defining $m^2 = (K/2)^2$. The quantum number $K$ is then restricted to even values by the requirement that $g(\gamma)$ be regular in the neighborhood of the origin. The solutions of Eq. (28) are

$$E_\gamma = \hbar\, \omega_\gamma (n_\gamma + 1), \qquad \omega_\gamma{}^2 = C_\gamma / B_\gamma, \tag{30a}$$

$$n_\gamma = 2N + |K|/2, \qquad N = 0, 1, 2, \ldots, \tag{30b}$$

$$g(\gamma) = \frac{N_\gamma}{\gamma} W_{(n_\gamma+1)/2,\, |K|/4} \left( \sqrt{B_\gamma C_\gamma}\; \gamma^2 / \hbar \right). \tag{30c}$$

Here $W_{k,m}(z)$ is a confluent hypergeometric function; the normalization constant $N_\gamma$ is determined by

$$\int_{-\infty}^{\infty} g^2(\gamma)\; |\gamma|\; d\gamma = 1.$$

The condition (30b) arises from the usual termination requirement on the solution (30c).

To the eigenvalues (29b) and (30a) must be added the rotational energy (22b) so that upon each beta or gamma level there is erected an appropriate rotational band. The ground state band has $K = 0$; and as the beta vibrations preserve the axial symmetry of the system, the beta band has the same value of $K$, for similar reasons, and has the spin sequence 0, 2, 4, ....

The first excited gamma vibrational band has but one quantum of excitation whence $|K| = 2$. Since the requirement on the $K$-quantum number is that $K \leq L$ the spin sequence of this band will be just $L = 2$, 3, 4, ....

There is no firm experimental evidence for bands with more than one quantum of excitation, although levels have been occasionally assigned values of $K > 2$. It is easy to see that the situation can become quite complicated; note that with $n_\gamma = 2$ there are two possible values for $K$: 0 and 4. Mixed beta-gamma bands are also possible. However, at these energies in even-even nuclei (about 2 MeV in the rare earths) the bands are not all collective; some of them are associated with the excitation of pairs from the core. The experimental determination of the collective-vs-pairing nature of these levels is not always clear cut; hence nothing further will be said concerning bands other than the three where $n_\beta = n_\gamma = 0$; $n_\beta = 1$, $n_\gamma = 0$; and $n_\beta = 0$, $n_\gamma = 1$. These are usually called the ground state, beta, and gamma bands respectively.

These bands all interact and influence one another; the interaction arises from the fact that $\mathcal{I}_1^{(2)}$ and $\mathcal{I}_2^{(2)}$ are not exactly equal. For small oscillations about the symmetric equilibrium shape, perturbation theory can be used to investigate the nature of this interaction. The perturbation Hamiltonian is just

$$H' = H - H^{\circ} = T_{\text{rot}} - T_{\text{rot}}^{\circ}, \tag{31}$$

where the superscript o refers to operators with $\mathcal{I}_1^{(2)} = \mathcal{I}_2^{(2)}$. By calling $\delta = \bar{\beta}/\beta_0$, (31) may be written to second order as

$$H' = \frac{\hbar^2}{2\mathcal{I}_0}\Big[(-2\delta + 3\delta^2 + 2\gamma^2)(L_1{}^2 + L_2{}^2) - \frac{2}{\sqrt{3}}(\gamma - 2\delta\gamma)(L_1{}^2 - L_2{}^2)\Big]. \tag{32}$$

Here the first term is diagonal in $L$ and $K$; but the second term, although diagonal in $L$, connects states with $\Delta K = 2$. Furthermore, because of the nature of the oscillator wave functions the term in $\delta$ will connect states where $\Delta n_\beta = 1$, $\delta^2$ states with $\Delta n_\beta = 0, 2$ and similarly for the terms in $\gamma$ and $\gamma^2$. That is, the $\delta$ term will cause an interaction between the beta and ground state bands; the $\gamma$ term will cause an interaction between the gamma and ground state bands. The diagonal contributions from the $\delta^2$ and $\gamma^2$ terms will lead to corrections in the moment of inertia $\mathcal{I}_0$; the off-diagonal contributions will be much smaller because of the energy denominator $\hbar\omega$. Finally, the $\delta\gamma$ term will cause an interaction between the beta and gamma bands; however, here the off-diagonal contribution can be quite large if the two band heads are close together (in $Th^{232}$, for instance, they are about 10 keV apart). By making use of the vibrational state functions given in Appendix C, it is a straightforward matter to obtain the diagonal contribution of $\delta^2$ and $\gamma^2$ and thus the first-order corrections $\Delta\mathcal{T}$ for the various bands are

$$\Delta\mathcal{T}_{\text{gnd}} = -\frac{\hbar}{2}\left(\frac{9}{\omega_\beta}+\frac{12}{\omega_\gamma}\right)$$

$$\Delta\mathcal{T}_{\beta} = -\frac{\hbar}{2}\left(\frac{27}{\omega_\beta}+\frac{12}{\omega_\gamma}\right)$$

$$\Delta\mathcal{T}_{\gamma} = -\frac{\hbar}{2}\left(\frac{9}{\omega_\beta}+\frac{24}{\omega_\gamma}\right).$$

Of greater interest are the level shifts caused by the band interactions. For example, the $\delta$ term of Eq. (32) has the matrix element between the beta and ground state bands.

$$\langle n_\beta = 0,\ \ n_\gamma = 0,\ \ L, M, K = 0 \mid \delta \mid\ n_\beta = 1,\ \ n_\gamma = 0,\ \ L, M, K = 0\rangle = \frac{-\hbar^2 L(L+1)}{\mathcal{T}_0\beta_0[2B_2\omega_\beta/\hbar]^{1/2}},$$

which in second order leads to a shift in each of the ground state levels by an amount

$$[\Delta E_{\text{gnd}}(L)]_\beta = -12\left(\frac{\hbar^2}{2\mathcal{T}_0}\right)^3 \frac{L^2(L+1)^2}{(\hbar\omega_\beta)^2},$$

which is identical with (20). Since, in second order, a level is pushed down by those above and up by those below, the ground state band will shift the levels of the beta band by an amount

$$[\Delta E_\beta(L)]_{\text{gnd}} = -[\Delta E_{\text{gnd}}]_\beta .$$

In a similar fashion, the shift in the ground state band due to the gamma band is found to be

$$[\Delta E_{\text{gnd}}(L)]_\gamma = -4\left(\frac{\hbar^2}{2\mathcal{T}_0}\right)^3 \frac{L^2(L+1)^2}{(\hbar\omega_\gamma)^2} \tag{33}$$

and, of course,

$$[\Delta_\gamma(L)]_{\text{gnd}} = -[\Delta E_{\text{gnd}}(L)]_\gamma.$$

Finally, the $\delta\gamma$ term of Eq. (32) leads to an energy shift

$$[\Delta_\beta E(L)]_\gamma = 48\left(\frac{\hbar^2}{2\mathscr{I}_0}\right)^4 \frac{L^2(L+1)^2}{[E_\beta(L) - E_\gamma(L)]\hbar\omega_\beta\hbar\omega_\gamma}. \tag{34}$$

This term will usually not be important because of the energy term $E_\beta(L) - E_\gamma(L)$ in the denominator. However, if it should happen that the beta and gamma levels of the same $L$ have almost the same energy, then clearly the contribution from Eq. (34) will be important. One case in which this is true is in $\text{Th}^{232}$ wherein the $\beta$ band $L = 2$ level is but 11 keV below the similar $\gamma$ band level. This difference is to be compared with the inertial constant $\hbar^2/2\mathscr{I}_0 = 8.29$ keV. (In $\text{Sm}^{152}$, for example, these values are 276 and 20.3 keV, respectively.)

These calculations have used second-order contributions (that is, the terms in $\delta^2$, $\gamma^2$, and $\delta\gamma$) in what is a first-order theory. In general this practice is permissible when the first-order contributions vanish (which is the case for electromagnetic $E2$ transitions between levels of negative parity; see Chapter 6, Section C) or when the second-order terms are relatively large (as in the case discussed before of the contributions arising from the $\delta\gamma$ term when $E_\beta(L) = E_\gamma(L)$). Some calculations have been carried to second order but not in a consistent manner [87]. To admit terms of order $|\alpha_{2\mu}|^2$ or $|a_{2\nu}|^2$ is permissible only if it can be shown that the contributions from the $\alpha_{4\mu}$ or $a_{4\nu}$ terms can be neglected. In the actinide region it has been shown that this is certainly not the case [116]; the magnitude of the $\alpha_{40}$ term is equal to or greater than the magnitude of the second-order terms $|\alpha_{2\mu}|^2$. Since this calculation has been based upon alpha-decay data these conclusions may not be applicable to the rare earth deformed region; however, caution should be exercised in using the results of any second-order calculation even in this region.

## C. Models without Axial Symmetry

In his original paper Bohr [56] developed the hydrodynamic model in a completely general fashion restricting the treatment to axial symmetry when he considered the coupling of the odd nucleon to the core. Marty first discussed the consequences in even-even nuclei of relaxing the requirement of axial symmetry [129]; Davydov and his co-workers have developed this model in greater detail.

The starting point is again the Hamiltonian of Eq. (3) with the moments of inertia of Eq. (16) but without assuming that the asymmetry parameter $\gamma$ be zero. In Appendix B it is shown that the general symmetry requirements impose the condition on the asymmetry parameter that its effective range be $0 \leq \gamma \leq \pi/3$, and in fact the Hamiltonian is symmetric about $\pi/6$. If we restrict $\beta$ to positive values, then $\gamma$ takes on the full range, giving prolate spheroids when $\gamma = 0$ and oblate spheroids when $\gamma = \pi/3$. Or, if we restrict the asymmetry parameter to the range $0 \leq \gamma \leq \pi/6$, then $\beta$ ranges over positive and negative values.

The simplest model assumption that can be made is that the nucleus is a deformed, asymmetric, rigid rotator [78]. Rigidity implies that $\dot{\beta} = \dot{\gamma} \equiv 0$ and since the nuclear surface possesses front-to-back symmetry, the eigenvalues belong to the $A$-representation. The calculation of the energy eigenvalues and eigenvectors is a straightforward matter and the eigenvalues (or their ratios to the lowest $L = 2$ state) have been tabulated [132] for values of the angular momentum less than 20. In Fig. 3-4 these eigenvalues are plotted as a function of $\gamma$ for $L \leq 12$. Note that for $\gamma = 0°(\pi/3)$ only even values of $L$ appear and the energy eigenvalues are given by Eq. (18). For other values of $\gamma$ relations (11a, c) are seen to be valid.

This simple model contains both the ground state and the gamma vibrational bands of the symmetric rotator theory. The band-head parameter $\hbar\,\omega_\gamma$ has been replaced by the asymmetry parameter $\gamma$. It was originally thought that the model would account for the low-lying structure of almost all even-even nuclei; however, it now appears to be useful only within the well-defined deformed regions. Its most obvious failing is that it cannot account for the beta-vibrational bands that are known to exist in deformed nuclei. This, of course, arises from the fact that there is no beta degree of freedom in the theory. We might hopefully add the effects of such a vibration by assuming it to be small and then treating the deviations from rigidity by perturbation theory. A calculation like this leads to an expression similar to Eq. (13b) where $b$ must be small compared with some mean value of the moments of inertia. Measurements of the energy that are associated with high spin values in even-even nuclei show that the magnitude of the moments of inertia change by a

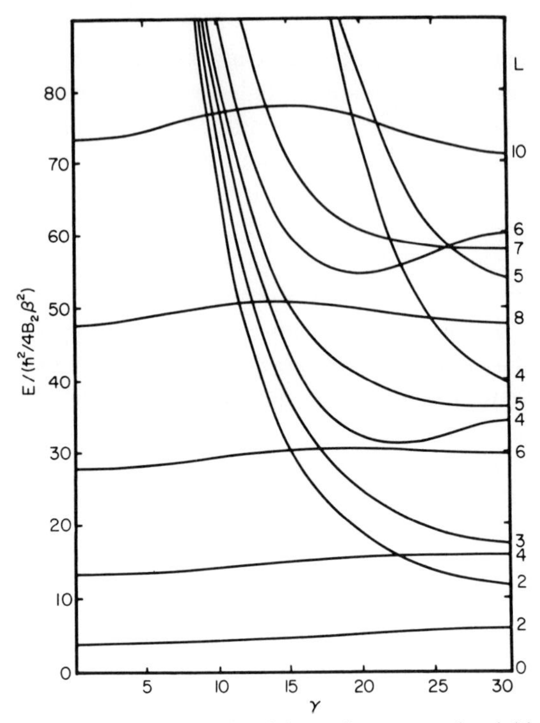


Fig. 3-4. The energy eigenvalues of a deformed, asymmetric, rigid rotator, that possesses hydrodynamic moments of inertia. The level structure is for the $A$-representation of the group $\mathbf{D}_2$ and is associated with the energy levels of even-even nuclei.

factor of more than two from the values given by the lowest spin values. Clearly these large vibrational effects cannot be accounted for by a perturbation calculation, and it is necessary to take these beta vibrations into account more exactly [79].

We can do this using the techniques of the last chapter, allowing a deformation ($\beta$) degree of freedom yet retaining the restriction that $\dot{\gamma} \equiv 0$. The Hamiltonian is then just

$$H = -\frac{\hbar^2}{2B_2}\left\{\frac{1}{\beta^3}\frac{d}{d\beta}\left(\beta^3\frac{d}{d\beta}\right) - \frac{1}{4\beta^2}\sum_{k=1}^{3}\frac{L_k{}^2}{i_k^{(2)}(\gamma)}\right\}$$
$$+ \frac{1}{2}C_2(\beta - \beta_0)^2. \tag{35}$$

Two important approximations are included here. They are the functional form of the moments of inertia and the vibrational potential. The irrotational form of the moments of inertia, proportional to $B_2\beta^2$, is used with the mass parameter $B_2$ held constant for each nucleus. The effect of centrifugal stretching is then accounted for by the quadratic dependence. It has been shown [83] that in the rare earth, deformed region this approximation is a reasonable one. No other functional form of the moments of inertia has been suggested. The second approximation is in the obvious extension of the potential function (2-22) to vibrations about a spheroidal equilibrium shape. The kinetic-energy operator was derived for vibrations about a spherical equilibrium shape for which the mass parameter is independent of $\mu$ (Eq. 2-21). Thus even though a contradiction exists in expression (35), it has been shown [83] that this form for the potential function is a close approximation to empirical-mass-formula potentials for moderate spins ($L < 16$).

The Schrödinger equation associated with this Hamiltonian separates into a rotational equation

$$\left[\frac{1}{2}\sum_{k=1}^{3}\frac{L_k{}^2}{i_k^{(2)}(\gamma)} - \mathscr{E}_{L,N}^{(2)}\right]\Psi_{L,N}^{(2)}(\theta_i, \gamma) = 0 \tag{36a}$$

and a vibrational equation

$$\left[-\frac{\hbar^2}{2B_2}\frac{1}{\beta^3}\frac{d}{d\beta}\left(\beta^3\frac{d}{d\beta}\right) + \frac{\hbar^2}{4B_2\beta^2}\mathscr{E}_{L,N}^{(2)} + \frac{1}{2}C_2(\beta-\beta_0)^2\right]\Phi_{L,Nn}^{(2)}(\beta) = E_{L,N,n}^{(2)}\Phi_{L,N,n}^{(2)}(\beta). \tag{36b}$$

The solutions of Eq. (36a) are simply those of a deformed asymmetric, rigid rotator and are displayed in Fig. 3-4.

Equation (36b) cannot be solved in closed form as it stands and a simplifying approximation must be made. The dependent variable is transformed by

$$\Phi_{L,N,n}^{(2)}(\beta) = \beta^{-3/2}D(\beta). \tag{37}$$

The differential equation now consists of a kinetic energy term and a generalized potential energy

$$W(\beta) = \frac{\hbar^2}{4B_2\beta^2}\left(\mathscr{E}_{L,N}^{(2)} + \frac{3}{2}\right) + \frac{1}{2}C_2(\beta-\beta_0)^2. \tag{38}$$

Here the first term, the centrifugal potential, gives rise to a new equilibrium deformation $\beta(L, N)$ that is the positive, real root of

$$\beta^4(L, N) - \beta^3(L, N)\beta_0 - \beta_0{}^4\mu^4(\mathscr{E}^{(2)}_{L,N} + 3/2)/2 = 0. \tag{39}$$

Here the stiffness parameter $\mu$ is related to the potential energy parameter $C_2$ by

$$\mu^2 = \hbar/\sqrt{C_2B_2}\,\beta_0{}^2. \tag{40}$$

By the variable transformation

$$y = Z_1\frac{\beta - \beta(L, N)}{\beta(L, N)}, \qquad -Z_1 \leq y \leq \infty$$

where

$$Z_1{}^4 = Z^4 + (3/2)(\mathscr{E}^{(2)}_{L,N} + 3/2)$$

and

$$Z = \beta(L, N)/\beta_0\mu.$$

Equation (36b) becomes

$$[d^2/dy^2 + 2\nu + 1 - y^2]D_\nu(\sqrt{2}y) = 0, \tag{41}$$

where the potential function (38) has been expanded about the new equilibrium deformation $\beta(L, N)$ and only the lowest nonvanishing term is kept. Equation (41) is subject to the boundary conditions

$$\begin{aligned} \lim_{y\to\infty} D_\nu(\sqrt{2}y) &= 0 \\ D_\nu(-\sqrt{2}Z_1) &= 0. \end{aligned} \tag{42}$$

If $\mu$ is equal to zero, then $Z_1$ becomes infinite and the solutions of Eq. (41) are the familiar ones of the linear harmonic oscillator. However, for $\mu$ *not zero* the second boundary condition determines the quantum number $\nu$ which is then not an integer. The energy eigenvalues are in any event

$$\begin{aligned} E^{(2)}_{L,N,n} = \hbar\,\omega_0\{&(\nu_n + 1/2)(Z_1/Z)^2 \\ &+ (\mathscr{E}^{(2)}_{L,N,n} + 3/2)\,[Z^4 + (\mathscr{E}^{(2)}_{L,N} + 3/2)/2]/4Z^6\}. \end{aligned} \tag{43}$$

The factor $\hbar\,\omega_0$ is an over-all scale parameter whereas the quantum number $n$ is the integer associated with the particular vibrational band: $n = 1$ for the ground state band, $n = 2$ for the beta-band, etc.

Equation (42) and from it, Eq. (43) must be solved numerically. A calculation such as this has been made with certain further approximations [160] although a computer program to calculate the $\nu_n$ is simple enough to write since the $D_\nu(z)$ are Weber's parabolic cylinder functions [20]. The boundary condition (42) therefore has the form

$$\frac{\Gamma(1/2)\, 2^{\nu/2} \exp(-Z_1^2/2)}{\Gamma(-\nu/2)\, \Gamma[(1-\nu)/2]} \sum_{k=0}^{\infty} \frac{(-1)^k\, \Gamma[(k-\nu)/2]}{k!} (2Z_1)^k = 0,$$

which is useful for numerical purposes.

Energy levels may now be labeled not only by their spin $L$ and the ordinal of that spin $N$ but also by $n$; thus we can write $E(LNn)$. The "gamma-band" head is then designated (221) whereas the beta-band head is designated by (012). An examination of the energy-level trends as a function of the stiffness parameter $\mu$ shows that as $\mu$ increases from zero each level is pushed down, the higher levels the most. However, no level crossings are generated by increasing $\mu$ within a given vibrational band $n$. Furthermore, the levels in the beta band decrease much more rapidly with increasing $\mu$ than do the ground state band levels. Since the levels in the "gamma band" decrease rapidly with increasing $\gamma$ and less rapidly with increasing $\mu$, these trends can be used to increase the accuracy of the root mean square deviation of experimental energy levels with theory. In practice the value of $\gamma$ is determined by the (221) level whereas the value of $\mu$ is obtained from the (012) or the (212) level. Then decreasing $\gamma$ and increasing $\mu$ (or vice versa) will hold the (221) level fixed and improve the fit. The results of several fitting procedures like these have been reported but apparently in no case has any attempt been made to obtain a "best fit" by varying $\gamma$ and $\mu$ to minimize the RMS deviation.

For small $\mu$, we can expand Eq. (43) in a power series in $\mathscr{E}^{(2)}_{LN_n}$ from which we can evaluate the parameter $b$ of Eq. (13b).

$$b = (1/2)\mu^4[1 + (57/4)(\nu + 1/2)\mu^2].$$

On dropping the restriction that $\dot{\gamma} \equiv 0$, that is, by permitting a $\gamma$ degree of freedom, the Schrödinger equation becomes

$$\left\{-\frac{\hbar^2}{2B_2}\left[\frac{1}{\beta^4}\frac{\partial}{\partial\beta}\left(\beta^4\frac{\partial}{\partial\beta}\right)+\frac{1}{\beta^2\sin^2 3\gamma}\frac{\partial}{\partial\gamma}\left(\sin 3\gamma\frac{\partial}{\partial\gamma}\right)\right.\right.$$

$$\left.\left.-\frac{1}{4\beta^2}\sum_{k=1}^{3}\frac{L_k^2}{\sin^2(\gamma-2\pi k/3)}\right]+V^B(\beta,\gamma)\right\}\Psi(\beta,\gamma,\theta_i)=E\Psi(\beta,\gamma,\theta_i). \quad (44)$$

The separation of this equation depends upon the choice of the potential function $V^B(\beta, \gamma)$. If the proper hydrodynamical potential is picked [Eq. (2-22), with $\lambda = 2$], then the system separates into the two equations

$$\left[\frac{-\hbar^2}{2B_2\beta^4}\frac{d}{d\beta}\left(\beta^4\frac{d}{d\beta}\right)+\frac{1}{2}C_2\beta^2+\hbar^2\Lambda/2B_2\beta^2\right]\Phi(\beta)=E\Phi(\beta) \quad (45a)$$

$$\left[-\frac{1}{\sin 3\gamma}\frac{d}{d\gamma}\left(\sin 3\gamma\frac{d}{d\gamma}\right)+\frac{1}{4}\sum_{k=1}^{3}\frac{L_k^2}{\sin^2(\gamma-2\pi k/3)}\right]\psi_L(\gamma,\theta_i)$$

$$=\Lambda\psi_L(\gamma,\theta_i). \quad (45b)$$

Since the Hamiltonian of this system and that of Eq. (2-10) are identical, the energy eigenvalues will be the same

$$E=\hbar\,\omega_0(N+5/2). \quad (46a)$$

The relation between the occupation number $N$ and $L$ has been given in Table 2-1 and it is only necessary to relate them to the separation parameter $\Lambda$. This is most easily seen by placing Eq. (45a) in the form

$$\left[-\frac{\hbar^2}{2B_2}\frac{d^2}{d\beta^2}+\frac{\hbar^2}{2B_2}\frac{(\Lambda+2)}{\beta^2}+\frac{1}{2}C_2\beta^2\right][\beta^2\Phi(\beta)]=E[\beta^2\Phi(\beta)] \quad (47a)$$

and then comparing this with the radial equation for the three-dimensional harmonic oscillator [6]

$$\left[-\frac{\hbar^2}{2M}\frac{d^2}{dr^2}+\frac{\hbar^2}{2M}\frac{l(l+1)}{r^2}+\frac{1}{2}kr^2\right]R_{nl}(r)=ER_{nl}(r). \quad (47b)$$

Replacing $l$ by a new parameter $(\lambda + 1)$ and comparing the second terms of Eq. (47a) and (47b) shows that

$$\Lambda=\lambda(\lambda+3).$$

The eigenvalue of Eq. (47b) is just

$$E_{nl}=\hbar\,\omega_0(2n'+l+3/2)=\hbar\,\omega_0(2n'+\lambda+5/2), \quad (46b)$$

which is to be equated to (46a) or to

$$N = 2n' + \lambda, \qquad n' = 0, 1, 2, \ldots.$$

By using this equation and Table 2-1, the quantities $N$, $L$, $\Lambda$, and $\lambda$ can be related. This is shown in Table 3-2.

TABLE 3–2

The Relation between Surfon Occupation Number $N$, Separation Parameter $\Lambda$, Total Angular Momentum $L$, and the Parameter $\lambda$ of Eq. (46b)

| $N$ | $\lambda$ | $\Lambda$ | $L$ |
|---|---|---|---|
| 0 | 0 | 0 | 0 |
| 1 | 1 | 4 | 2 |
| 2 | 0 | 0 | 0 |
| | 2 | 10 | 2, 4 |
| 3 | 1 | 4 | 2 |
| | 3 | 18 | 0, 3, 4, 6 |
| 4 | 0 | 0 | 0 |
| | 2 | 10 | 2, 4 |
| | 4 | 28 | 2, 4, 5, 6, 8 |
| . | . | . | . |
| . | . | . | . |
| . | . | . | . |

The eigenfunctions $\psi_L(\gamma, \theta_i)$ have a very simple form for $L=0, 3$:

$$\psi_0(\gamma, \theta_i) = N_0 D^0_{00}(\theta_i) P_{\lambda/3} (\cos 3\gamma)$$
$$\psi_3(\gamma, \theta_i) = N_3[D^3_{M2}(\theta_i) - D^3_{M-2}(\theta_i)] P^1_{\lambda/3} (\cos 3\gamma),$$

where $P_n$ is a Legendre function $P_n{}^m$ is the associated function, and $N_l$ is a normalization constant. Solutions for values of $L \leq 6$ have been reported [112]. However, they have not proved very useful for deformed nuclei since the spectrum, Eq. (46a), bears little resemblance to the level structure of these nuclei.

This level structure is due to the angular momentum degeneracy of the solutions which stems from the form of the potential $V(\beta, \gamma)$. In order to remove this degeneracy the potential must be changed; however, these are conflicting requirements on the potential. In order to carry out the separation of $\beta$ and $\gamma$ variables as was done in Eqs. (45a) and (45b), the potential must have the form

$$V^B(\beta, \gamma) = (1/2)C_2\beta_2 + (1/\beta^2/f(\gamma).$$

The form this potential takes from a second-order hydrodynamical calculation is

$$V^B(\beta, \gamma) = (1/2)C_2\beta^2(1 + \sin^2\gamma),$$

which will not permit separation.

A recipe for an approximate separation has been given by Davydov [80] in which a gamma potential

$$V(\gamma) = \varkappa(\gamma - \gamma_0)^2$$

is added to Eq. (45b) *after* separation. He investigated gamma vibrations of spherical nuclei, symmetric deformed nuclei, and asymmetric deformed nuclei. It was necessary to use a different approximation for each type of system. Only a very few comparisons with experiment have been made with this model, and from the nature of the approximations, these comparisons are probably not too enlightening. One point Davydov does make is that the model predicts the existence of two excited 0+ states: one associated with the $\beta$-vibrational band and the second associated with the $\gamma$ vibrations. This second 0+ state is sometimes termed the "double-gamma" band head. However, great caution must be exercised in assuming that such second 0+ states, where they exist, are indeed to be associated with a collective gamma vibration. Gallagher and Soloviev [30] have shown that at about the same energy in even-even nuclei (about 1 MeV in the rare earth region) there are states of a strong "quasiparticle" nature.

## D. Theories of Odd-Parity States

Even-even nuclei throughout the periodic table are found to have negative parity states. In the deformed regions in which they have been observed they do not form a vibrational-type spectrum (Fig. 2-1, $\lambda = 3$). The lowest negative parity states occur at about 1 MeV in the rare earth deformed region and from 2–500 keV in the actinide region. The suggestion that these states probably arise from an octupole ($\lambda = 3$) deformation of the nuclear surface was made quite early [23].

Again two types of theories have been developed to account for these levels. One type treats octupole vibrations about either a spherical [117] or a spheroidal [123] equilibrium shape, whereas the other supposes that these levels arise from the rotations and vibrations of a pure octupole-shaped surface [72]. These types will be discussed in turn.

Qualitatively, if the nuclear surface can undergo both quadrupole ($\pi+$) and octupole ($\pi-$) oscillations simultaneously, then the positive parity level structure will be essentially the same as that discussed before. However, the lowest negative parity state will not be the 3— state but a 1— level arising from the coupling of the 3— with a 2+ level. If the oscillations are about a spheroidal equilibrium shape and are essentially symmetric, then the bands that arise will be associated, as before, only with even values of $K$. Since 1 — levels are observed in many deformed even nuclei the discussion here will be limited to oscillations about a spheroidal equilibrium shape.

The moments of inertia for $\lambda = 3$ can be written in terms of the body-fixed, octupole expansion parameters as

$$\begin{aligned} \mathscr{I}_1^{(3)} &= B_3(6a_{30}^2 + 2\sqrt{30}\, a_{30}a_{32} + 8a_{32}^2) \\ \mathscr{I}_2^{(3)} &= B_3(6a_{30}^2 - 2\sqrt{30}\, a_{30}a_{32} + 8a_{32}^2) \\ \mathscr{I}_3^{(3)} &= 8B_3a_{32}^2\,, \end{aligned} \tag{48}$$

where the inertial tensor has been diagonalized by setting the $a_{3\pm\mu} = 0$ for $\mu = 1$, 3. (As we discussed in the last chapter an alternative way to diagonalize the inertial tensor is to set the $a_{3\pm\mu} = 0$ for $\mu$ even. The resulting moments of inertia have been given [123]. However, Soloviev et al. [147] have shown in a quasiparticle calculation of these negative parity states that the $\mu = \pm 1, \ \pm 3$, degrees of freedom are not, in general, of a collective nature; thus by setting the $a_{3,\mu} = 0$, $\mu$ odd is physically reasonable in this collective model discussion.

The classical kinetic energy for oscillations about a spheroidal equilibrium shape is then

$$T_{\text{vib}} = (1/2)B_2(\dot{\beta}^2 + \beta^2\dot{\gamma}^2) + (1/2)B_3(\dot{a}_{30}^2 + 2\dot{a}_{32}^2).$$

If the octupole oscillations preserve axial symmetry, then the relation involving the 1 and 2 components of the moments of inertia of Eq. (21) are unchanged. However the third component becomes

$$\mathscr{I}_3^{(2+3)} \cong 4B_2B_0{}^2\gamma^2 + 8B_3a_{32}^2 = 4B_2\beta_0{}^2(\gamma^2 + g^2). \tag{49}$$

The complete system is quantized as before (Section B) by placing the rotational term

$$\hbar^2 K^2/2\mathscr{I}_3^{(2+3)}$$

in the vibrational part of the Hamiltonian

$$\begin{aligned} H_{\text{vib}} = &-\frac{\hbar^2}{2}\left\{\frac{1}{B_2}\frac{\partial^2}{\partial\beta^2} + \frac{1}{B_3}\frac{\partial^2}{\partial b^2}\right. \\ &\left. + \frac{1}{B_\gamma}\left[\frac{\partial^2}{\partial\gamma^2} + \frac{\gamma}{\gamma^2+g^2}\frac{\partial}{\partial\gamma} + \frac{\partial^2}{\partial g^2} + \frac{g}{\gamma^2+g^2}\frac{\partial}{\partial g}\right]\right\} \\ &+ \frac{(\hbar K/2)^2}{2B_\gamma(\gamma^2+g^2)} + V(\beta, b, \gamma, g)\,, \end{aligned} \tag{50}$$

where $B_\gamma = B_2\beta_0{}^2$ and $b = a_{30}$. The potential energy is assumed to have the harmonic form

$$V(\beta, b, \gamma, g) = (1/2)C_2(\beta - \beta_0)^2 + (1/2)C_3b^2 + (1/2)C_\gamma\gamma^2 + (1/2)C_g g^2.$$

From the form of this potential function (which is not consistent with a hydrodynamical calculation), the $\beta$ and $b$ variables separate from the vibrational Schrödinger equation yielding two equations identical with (27), one has $\beta$ as the variable and the other, $b$. The $\beta$ solutions are identical with Eq. (29a) and (29b) and the $b$ solutions are similar:

$$\begin{aligned} f(b) &= N_bH_{n_b}(\sqrt{B_3\omega_b/\hbar}\, b)\exp(-\, B_3\omega_b b^2/2\hbar) \\ E_b &= \hbar\,\omega_b(n_b + 1/2), \qquad n_b = 0,\ 1,\ 2,\ \ldots. \end{aligned}$$

The normalization is determined by a condition identical with (29c).

The remaining equation in $\gamma$ and $g$ can only be approximately separated by introducing new variables

$$\begin{aligned} \gamma &= \Gamma\cos\sigma \\ g &= \Gamma\sin\sigma\,, \end{aligned}$$

defined on the interval

$$0 \leq \Gamma < \infty$$
$$0 \leq \sigma \leq 2\pi ,$$

and assuming that

$$C_g = C_\gamma + \delta .$$

If $\delta$ is small compared with $C_\gamma$ the separation of $\Gamma$ and $\sigma$ may be carried out with the perturbation term

$$H_1 = (1/2)\delta(\Gamma \sin \sigma)^2. \tag{51}$$

The zeroth-order state functions are

$$\psi(0) = P(\Gamma)\, S(\sigma)$$

and satisfy

$$\left\{\frac{-\hbar^2}{2B_\gamma}\left[\frac{1}{\Gamma^2}\frac{d}{d\Gamma}\left(\Gamma^2\frac{d}{d\Gamma}\right) - \frac{\mu^2 + (K/2)^2}{\Gamma^2}\right] + \frac{1}{2}C_\gamma \Gamma^2\right\} P(\Gamma) = E_\Gamma P(\Gamma) \tag{52a}$$

$$\frac{d^2 S}{d\sigma^2} + \mu^2 S = 0. \tag{52b}$$

Appropriate boundary conditions on Eq. (52b) require $\mu$ to be integral; then setting

$$\mu^2 + (K/2)^2 = \lambda(\lambda + 1),$$

Eq. (52a) is identical with the radial equation of the three-dimensional oscillator with solutions

$$\begin{aligned} E_\Gamma &= \hbar\sqrt{C_\gamma/B_\gamma}\,(n_\Gamma + 3/2) \\ n_\Gamma &= 2k + \lambda, \qquad k = 0,\ 1,\ 2,\ \ldots. \end{aligned} \tag{53}$$

The state with one quantum of excitation has $n = 1$ whence $k = 0$, $\lambda = 1$, and $|K| = 2$. The state is doubly degenerate in the $\mu$ quantum number. This degeneracy is removed in first order by the perturbation of Eq. (51), yielding the first-order energy corrections

$$E_1^{(1)} = \delta\, \hbar/4\sqrt{B_\gamma C_\gamma} \tag{54a}$$

$$E_2^{(1)} = 3\delta\, \hbar/4\sqrt{B_\gamma C_\gamma}. \tag{54b}$$

Thus there are two excited bands with $|K| = 2$ and $I = 2, 3, 4, \ldots$: the $\gamma$ band has even parity whereas the $g$ band has odd parity. If $\delta > 0$ the $\gamma$ band lies below the $g$ band whereas if $\delta < 0$ the opposite is true. From Eq. (54) it is seen that the band heads are separated by

$$|\Delta E| = |\delta|\, \hbar/2\sqrt{B_\gamma C_\gamma}.$$

The $\gamma$ and $g$ bands can be relatively close together. In the nucleus $W^{182}$ the 2+ and 2— levels are about 40 keV apart with the band heads being at 1258 and 1290 keV respectively.

It has been pointed out [121] that this treatment assumes that the principal axes for the quadrupole and octupole vibrations are and remain coincident. The theory can be suitably changed to remedy the need for this assumption by the introduction of further parameters.

The other model of interest is that of a rotating, pure octupole shape with superimposed deformation ($\beta$-like) vibrations. Since the rotator need not have an axis of symmetry this model is similar to that developed in Section C. The moments of inertia are parameterized using Eq. (2-19) and setting $\beta_3 = \zeta$. The most general expansion is given by

$$\begin{aligned} \varepsilon_{30} &= \cos\eta\ \cos\iota \\ \varepsilon_{3\pm1} &= (\pm 1/\sqrt{2})\cos\xi\ \sin\iota \\ \varepsilon_{3\pm2} &= (1/\sqrt{2})\sin\eta\ \cos\iota \\ \varepsilon_{3\pm3} &= (\pm 1/\sqrt{2})\sin\xi\ \sin\iota \end{aligned} \tag{55}$$

where $\iota$, $\eta$, and $\xi$ are generalized asymmetry parameters of the octupole surface. The usual treatment of negative-parity systems is to set $\iota = 0$. This diagonalizes the inertial tensor but removes two degrees of freedom from the system. For this circumstance, the moments of inertia are

$$\begin{aligned} \mathcal{I}_1^{(3)} &= 4B_3\zeta^2[\sin^2\eta + (1/2)\sqrt{15}\sin\eta\ \cos\eta + (3/2)\cos^2\eta] \\ \mathcal{I}_2^{(3)} &= 4B_3\zeta^2[\sin^2\eta - (1/2)\sqrt{15}\sin\eta\ \cos\eta + (3/2)\cos^2\eta] \\ \mathcal{I}_3^{(3)} &= 4B_3\zeta^2\sin^2\eta. \end{aligned} \tag{56}$$

Because the nuclear surface does not have "front-to-back" symmetry for these oscillations, the appropriate representation of the symmetry class $\mathbf{D}_2$ to be used is $B_1$ [72]. The asymmetry parameter $\eta$ has the range $0 \leq \eta \leq \pi$ with the eigenvalues being symmetric about $\pi/2$. These eigenvalues are obtained by the usual diagonalization of the Hamiltonian of Eq. (3) in the representation (5). The eigenvalues have been tabulated for $L \leq 12$ as a function of $\eta$ ($\Delta\eta = 5°$) [160].

Deformation ($\zeta$ —) vibrations are added in exactly the same manner to this rotating, pure octupole model as they were added to the asymmetric rotator model of the quadrupole surface (Section C); and in fact it has been shown that deformation vibrations can be added to any rotator whose surface can be represented by a single-order spherical harmonic [160].

These models cannot describe negative parity levels belonging to odd values of $K$. Although bands such as these have not been definitely found, their existence would require either using the $a_{3,\mu}$, $\mu = \pm 1, \pm 3$ expansion parameters in the above theories, or dropping the restriction that $\iota = 0$ in Eq. (55). The latter program has been carried out [76] but the resulting theory is by far too complicated to be applied unambiguously to the rather sparse collection of negative parity data now in existence.

CHAPTER 4

# Models of Odd-$A$ Nuclei

## A. The Coupling of a Single Particle and the Core

The extension of these collective models to odd-$A$ and odd-odd-$A$ nuclear systems assumes that an odd number of protons (and/or neutrons) is coupled to an even-even core. The most usual assumption is to take the core to consist of all of the nucleons but one (or two); however, we could as well assume the core to consist of the nearest doubly magic nucleus to which are coupled the proper number of particles, holes or both to constitute the given system. Calculations such as these will be considerably more complex so that in what follows the discussion will be restricted to the coupling of a single particle to a deformed even-even core. In the next chapter a short discussion will be given of odd-odd-$A$ nuclei in which an odd neutron and an odd proton are coupled to this type of deformed core.

Furthermore, the treatment will be restricted to strong coupling models for which it is assumed that the motion of the odd nucleon in the potential formed by the core and seen from the reference frame fixed in the nucleus is identical with that of a nucleon moving in a similar, but spatially fixed, potential. The question of the exact form this potential has can be deferred until later, but the assumption of strong coupling effects the commutators of the angular momentum operators. Let $\mathbf{L}$ now be the angular momentum of the core and $\mathbf{j}$, the angulai momentum of the extra core nucleon. Because of the deformation neither of these will be constants of the motion; however, the total nuclear angular momentum $\mathbf{I}$ will be

$$\mathbf{I} = \mathbf{L} + \mathbf{j}. \tag{1}$$

Then the assumption of strong coupling implies that in the body-fixed reference frame

$$[I_1, I_2] = -iI_3\,, \text{ cyclically}$$

$$[j_1, j_2] = ij_3\,, \text{ cyclically}.$$

The projection of **I** upon the laboratory $Z$ axis is $M$; its projection on the body-fixed 3 axis is taken as $K$ whereas the projection of **j** on this axis will be denoted by $\Omega$. These vectors and their projections are shown schematically in Fig. 4-1.

For the most general deformation only $\mathbf{I}^2$ and $I_z$ will be constants of the motion. If the system possesses an axis of symmetry, then $K$ and $\Omega$ will also be good quantum numbers. In such a case, the axis of symmetry

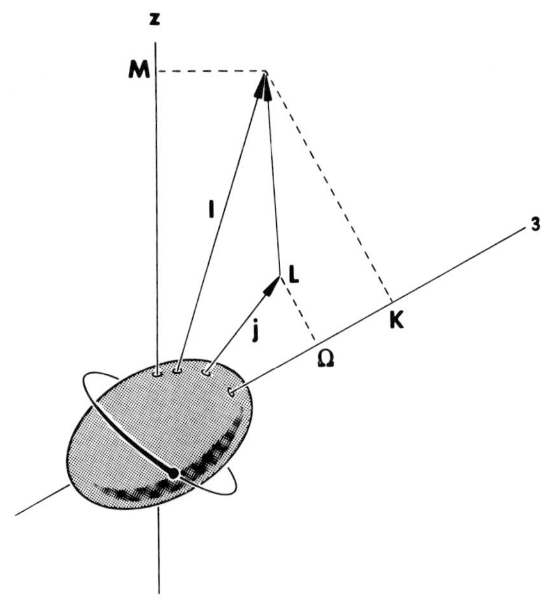


Fig. 4-1. The projection of the total angular momentum **I** of an odd-$A$ nucleus on the laboratory $Z$ axis ($M$) and on the body-fixed 3 axis ($K$). The quantum number $\Omega$ is the projection of the odd-nucleon angular momentum **j** on this latter axis. Only if the nucleus possesses axial symmetry will $K$ and $\Omega$ be constants of the motion. As long as there is any nuclear deformation, $\mathbf{j}^2$ will not be a constant of the motion.

will be taken as the 3 axis so that $K$ and $\Omega$ will be the appropriate projections on the symmetry axis.

If the nucleus is symmetric it is reasonable to suppose that the core will be too and therefore will behave as a similarly deformed even-even nucleus. In this case, at least for the ground state rotational band, $L_3 = 0$; thus for the odd-$A$ case, Eq. (1) shows that $K = \Omega$. The Hamiltonian can now be constructed, for the lowest levels, and is the sum of the corc kinetic energy and the Hamiltonian of the odd nucleon

$$H = H_{\text{rot}} + H_{\text{part}} = \frac{\hbar^2}{2} \sum_{k=1}^{3} \frac{L_k{}^2}{\mathscr{I}_k} + \left(\frac{p^2}{2m} + V\right).$$

The potential function $V$ is not only a function of the coordinates $\mathbf{r}$ but possibly also of the nucleon angular momentum $\mathbf{l}$ and the nucleon spin $\mathbf{s}$ as is the case with the shell model potential to which $V$ must return when the deformation vanishes. If one makes use of Eq. (1) and the condition of axial symmetry $\mathscr{I}_1 = \mathscr{I}_2 = \mathscr{I}_0$, then this Hamiltonian can be written as the sum of three terms

$$H = H_R + H_p + H_c,$$

where

$$H_R = \frac{\hbar^2}{2\mathscr{I}_0} [I(I+1) - 2K^2] \tag{2}$$

$$H_p = \frac{p^2}{2m} + V(\mathbf{r}, \mathbf{l}, \mathbf{s}) + \frac{\hbar^2}{2\mathscr{I}_0} \mathbf{j}^2 \tag{3}$$

$$H_c = -\frac{\hbar^2}{\mathscr{I}_0} (I_1 j_1 + I_2 j_2). \tag{4a}$$

Since this last term is of the form $\boldsymbol{\omega} \cdot \mathbf{j}$ that classically would give rise to the Coriolis force, it will be called the Coriolis term (the designation RPC, rotation–particle coupling, is also used). The effect of applying $H_c$ is most easily seen by rewriting it with respect to the stepping operators, where for any vector operator $v$

$$v^{\pm} = v_1 \pm i v_2$$

so that (4a) becomes

$$H_c = \frac{-\hbar^2}{2\mathscr{I}_0} (I^+ j^- + I^- j^+). \tag{4b}$$

The first term steps both $K$ and $\Omega$ down whereas the second term steps them both up. In general then, $H_c$ possesses only off-diagonal matrix elements except for the case where $K = \Omega = \frac{1}{2}$ for which the nonvanishing matrix elements of $H_c$ are diagonal [71]. This follows from the fact that since the nucleus is assumed to possess axial symmetry the levels are degenerate in the sign of $K$ and $\Omega$. We shall return to this issue shortly since it is more illuminating to discuss the various approximations that are used in solving the Schrödinger equation for these odd-$A$ systems.

If the particle is very tightly bound to the core, then the particle level spacings will be large compared with the rotational level spacings, thus, aside from the single case in which there are diagonal calculations, we may neglect $H_c$. The Schrödinger equation is then

$$H \mid EIMK = \Omega\rangle = (H_R + H_p) \mid EIMK = \Omega\rangle = E_{I,K} \mid EIMK = \Omega\rangle,$$

which can be separated: the state functions are the product of a rotational function $\mid EIMK\rangle$, that is associated with the eigenvalue problem

$$H_R \mid EIMK\rangle = \frac{\hbar^2}{2\mathscr{I}_0} [I(I+1) - 2K^2] \mid EIMK\rangle ,$$

and a particle function $\mid \Omega\rangle$ that is the solution of

$$H_p \mid \Omega\rangle = E_\Omega \mid \Omega\rangle .$$

Thus

$$E_{I,K} = \frac{\hbar^2}{2\mathscr{I}_0} [I(I+1) - 2K^2] + E_\Omega . \tag{5}$$

Clearly, $I \geq K$; therefore in this approximation the level structure is that of rotational bands with sequence $I = K = \Omega,\ K+1,\ K+2,\ \ldots,$ built upon the particle levels $E_\Omega$.

The properly symmetrized state functions can be easily constructed by using properties of the transformation operators $T_1$ and $T_2$ of Appendix A once the particle state functions $\mid \Omega\rangle$ are expanded in terms of a complete set of isotropic solutions $|j\Omega\rangle$. That is, if

$$\mid \Omega\rangle = \sum_j C_{j\Omega} \mid j\Omega\rangle ,$$

then

$$| EIMK\rangle = \frac{1}{\sqrt{2}} \sum_j C_{j\Omega}[D^I_{MK}(\theta_i) \mid jK\rangle + (-1)^{I-j} D^I_{M-K}(\theta_i) | j - K\rangle] \tag{6}$$

The relation

$$| j\Omega\rangle = \sum_{\Omega'} D^i_{\Omega'\Omega} \mid j\Omega'\rangle_L ,$$

where $|j\Omega'\rangle_L$, the solution of the isotropic problem that is referred to the laboratory frame, is useful here.

Finally, what is measured, of course, is the level sequence and spacings which are given by

$$\Delta E(I, K) = E_{IK} - E_{KK} = \frac{\hbar^2}{2\mathscr{I}_0} [I(I+1) - K(K+1)]. \tag{7a}$$

A good example of this simple level structure is to be found in the odd-proton nucleus $Tb^{159}$, the lowest rotational band of which has $K = \frac{3}{2}$. Six excited states with $I \leq 15/2$ have been definitely assigned to this band. The average moment of inertia parameter $\hbar^2/2\mathscr{I}_0$ is 11.35 keV with an RMS deviation of less than 4%. The assumption that the particle levels are widely separated when compared with the rotational levels is fairly well met as the nearest known particle state is at 348 keV (compared with 58.0 keV for the $I = \frac{5}{2}+$, $K = \frac{3}{2}$ rotational level) and has $K = \frac{5}{2}+$. This level sequence is seen in Fig. 4-2. Many nuclei have one or more rotational bands whose level structure agrees well with these predictions that do not require the solution to the particle-particle problem $E_\Omega$.

As mentioned earlier, $H_c$ possesses diagonal contributions for the case when $K = \frac{1}{2}$ and despite the strength of the core-particle coupling these contributions should be included in the expression (7a). For symmetric nuclei the most general expression for the matrix elements of $H_c$ is

$$\langle EIMK' = \Omega' \mid H_c \mid EIMK = \Omega\rangle =$$

$$= -\frac{\hbar^2}{2\mathscr{I}_0} \sum_j C^*_{j\Omega'} C_{j\Omega} \{[\delta_{K', K-1}$$

$$+ (-1)^{I-j} \delta_{K', -(K-1)}][(I+K)(I-K+1)(j+K)(j-K+1)]^{1/2}$$

$$+ [(I-K)(I+K+1)(j-K)(j+K+1)]^{1/2} \delta_{K', K+1}\}. \tag{8}$$

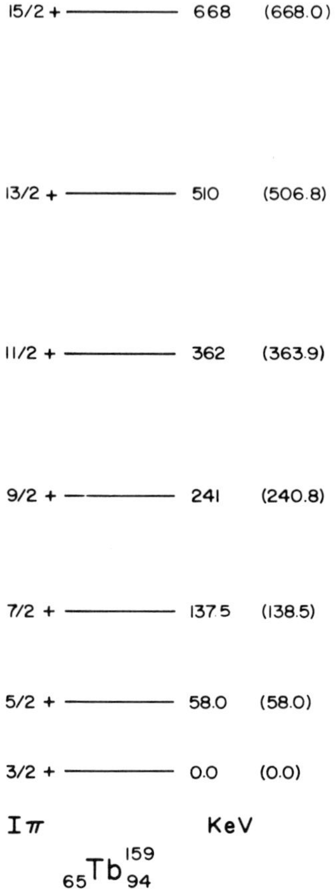

| $I\pi$ | KeV | |
|---|---|---|
| 15/2 + | 668 | (668.0) |
| 13/2 + | 510 | (506.8) |
| 11/2 + | 362 | (363.9) |
| 9/2 + | 241 | (240.8) |
| 7/2 + | 137.5 | (138.5) |
| 5/2 + | 58.0 | (58.0) |
| 3/2 + | 0.0 | (0.0) |

$_{65}\mathrm{Tb}^{159}_{94}$

Fig. 4-2. The $K = \frac{3}{2}+$ rotational band in the odd-proton nucleus $\mathrm{Tb}^{159}$. A mean inertial parameter of 11.35 keV is obtained from Eq. (7a). The values in parentheses are calculated from Eq. (7c) with the parameter values $\hbar^2/2\,\mathscr{I}_0 = 11.63$ keV and $b = 8.48 \times 10^{-3}$ keV.

When $K = \frac{1}{2}$ there is a diagonal contribution from the second term and this is usually written as

$$\langle EIM\,\tfrac{1}{2} \mid H_c \mid EIM\,\tfrac{1}{2}\rangle = (-1)^{I+1/2}\,\frac{\hbar^2}{2\mathscr{I}_0}\,(I + \tfrac{1}{2})a$$

which also serves to define the "decoupling parameter" $a$

$$a = \sum_j (-1)^{j+1/2}\,(j + \tfrac{1}{2}) \mid C_{j1/2} \mid^2. \tag{8a}$$

These diagonal contributions are to be added to the level spacing relation (7a)

$$\Delta E(I, K) = \hbar^2/2\mathcal{J}_0 \{I(I+1) - K(K+1) + a[(-1)^{I+1/2}(I+\tfrac{1}{2}) + 1]\, \delta_{K,1/2}\}. \tag{7b}$$

The decoupling parameter can have a very strong influence on the appearance of a $K = \frac{1}{2}$ rotational band. If it is very small, the band will display the usual $I(I+1)$ spacing. For larger values, because of the alternating behavior of the sign, the level spacing may appear as a number of closely spaced doublets. And if $a$ is sufficiently large and negative,

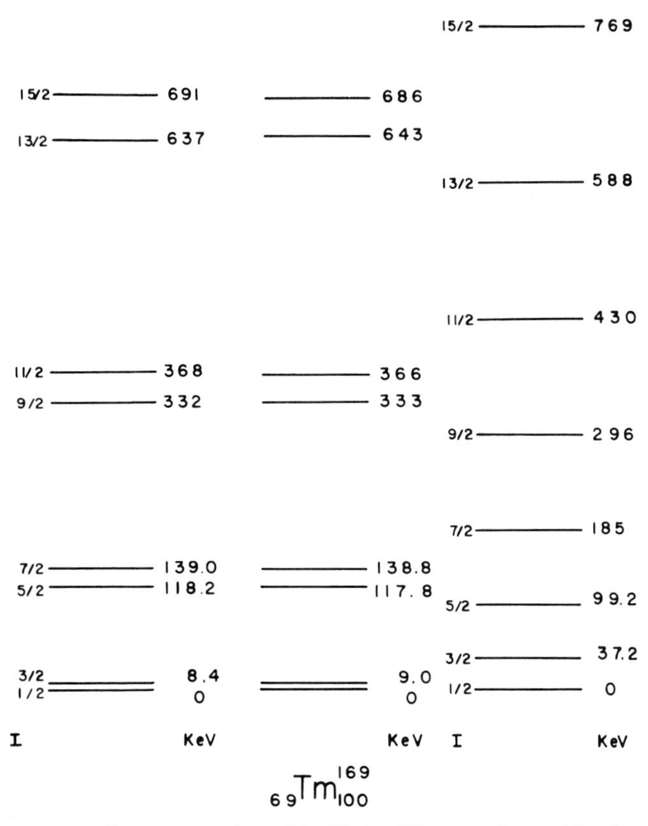


Fig. 4-3. The $K = \frac{1}{2}+$ rotational band in $Tm^{169}$. The experimental levels are to the left. In the center are the levels calculated from the appropriate Equation (7) with parameter values $\hbar^2/2\mathcal{J}_0 = 12.43$ keV, $b = 3.578 \times 10^{-3}$ keV, $a = -0.7575$. On the far right are the levels that are obtained by setting $a$ equal to zero.

the band will have an anomalous behavior: its $I = \frac{3}{2}$ level will appear *below* the $I = \frac{1}{2}$ level. The importance the decoupling parameter can have on this type of band structure is found, for example, in the odd-proton nucleus $Tm^{169}$ wherein at least eight levels with $I < 15/2$ are known to belong to the $K = \frac{1}{2}$ ground state band. The observed level sequence is normal $E(I, \frac{1}{2}) < E(I + 1, \frac{1}{2})$ but the spacings cannot be accounted for by the simple scheme of Eq. (7a) since the levels are grouped into four relatively close doublets. The addition of the diagonal contributions of $H_c$ brings the calculated spacings into very close agreement with the measured values. This yields a value of the inertial parameter of 12.38 keV; the decoupling parameter is found to be $-$ 0.77. These values give an RMS error of less than 1%, which is remarkable in view of the very elementary model being used. In Fig. 4-3 the measured energies for this nucleus are compared with those calculated from Eq. (7b) using the parameter values given here.

At this point it is possible to include the effect of vibrational degrees of freedom. If these are of sufficiently high frequency, then they may be treated by perturbation methods as for even nuclei by including a stretching parameter $b$ in Eq. (7a) or Eq. (7b). Thus the equation becomes

$$\Delta E(I, K) = \frac{\hbar^2}{2\mathscr{I}_0}[I(I+1) - K(K+1)] \\ - b\{[I+1) - 2K^2]^2 - K^2(K-1)^2\} \tag{7c}$$

with an obvious extension to $K = \frac{1}{2}$ bands. For the example $Tb^{159}$ the inclusion of this effect leads to the values $\hbar^2/2\mathscr{I}_0 = 11.63$ keV and $b = 8.48 \times 10^{-3}$ keV which reduces the RMS error to 2.2%. For $Tm^{169}$ the three parameters become $\hbar^2/2\mathscr{I}_0 = 11.97$ keV, $a = -0.77$, and $b = 3.4 \times 10^{-2}$ keV. For this nucleus the nearest particle level that could perturb this simple structure is at 316 keV and is the head of a $K = 7/2+$ band.

In order to account for this perturbation by other particle states we make use of the general expression (8) to form the energy matrix from which we extract the secular determinant

$$\begin{vmatrix} E(I,K) - E(I) & \langle K \mid H_c \mid K+1\rangle & 0 & \cdots \\ \langle K+1 \mid H_c \mid K\rangle & E(I,K+1) - E(I) & \langle K+1 \mid H_c \mid K+2\rangle & \cdots \\ 0 & \langle K+2 \mid H_c \mid K+1\rangle & E(I,K+2) - E(I) & \cdots \\ \vdots & \vdots & \vdots & \vdots \end{vmatrix} = 0.$$

There are a number of cases in which two adjacent bands of the same parity with $\Delta K = 1$ exist in odd-$A$ nuclei whose interaction is easily calculated. The solutions of the secular equation for this case are

$$E(I) = \frac{1}{2} \{\Sigma E \pm \delta E[1 + (I - K)(I + K + 1) \mid 2A_K/\delta E \mid^2]^{1/2}\}, \quad (9)$$

where

$$\begin{aligned} \Sigma E &= E(I, K) + E(I, K + 1) \\ \delta E &= E(I, K) - E(I, K + 1) \\ A_K &= - \hbar^2/2\mathcal{I}_0 \sum_j C^*_{jK+1} C_{jK}[(j - K)(j + K + 1)]^{1/2} \end{aligned} \quad (9a)$$

This last factor can be calculated once the solutions to the particle problem are known; however, it is usually obtained by fitting (9) to the experimental level scheme.

It is clear from Eq. (9) that $K$ is no longer a good quantum number since each level in this example is a mixture of two. This $K$-band superposition is essentially equivalent to starting with slight asymmetries. If the particle is but slightly coupled to the core then the radicand will differ but little from unity and $K$ will be an "approximately good" quantum number. It is illuminating, under these circumstances to expand Eq. (9) to lowest order. The corrected $K$-band energies are then

$$E'(I, K) = E(I, K) + \frac{(I - K)(I + K + 1)A_K^2}{\delta E} \quad (10a)$$

$$E'(I, K + 1) = E(I, K + 1) - \frac{(I - K)(I + K + 1)}{\delta E} A_K^2. \quad (10b)$$

The level spacing can now be written in the form of Eq. (7a) with an effective inertial parameter $\hbar^2/2\mathcal{I}_0(K)$, where

$$\frac{\hbar^2}{2\mathcal{I}_0(K^*)} = \frac{\hbar^2}{2\mathcal{I}_0} \pm \frac{A_K^2}{\delta E}.$$

The upper sign is associated with $K^* = K$; the lower sign, with $K^* = K + 1$. If the bands are such that the lower energy band has the lower value of $K$ whereas the upper band has the value $K + 1$ (so that $\delta E < 0$), then the effective moment of inertia of the lower band is increased whereas that of the upper band is decreased. If the order of the bands is reversed, then the effect on the moments of inertia $\mathcal{I}_0(K^*)$ is just the opposite of this.

Sometimes the expression for the energy levels in nuclei with admixed $K = \frac{1}{2}$ and $\frac{3}{2}$ bands is given in the form

$$E_I = E_0 + AI(I+1) + BI^2(I+1)^2 + C(-1)^{I+1/2}(I-\tfrac{1}{2})(I+\tfrac{1}{2})(I+\tfrac{3}{2}), \tag{10c}$$

where the constant $C$ is written as

$$C = \frac{a\hbar^2}{2\mathcal{I}_0}\left|\frac{\hbar^2\langle\frac{3}{2}\,|j^-\,|\,\frac{1}{2}\rangle}{2\mathcal{I}_0\,\delta E}\right|^2.$$

Although this is not obviously of the form of Eq. (10a), it can also be obtained from Eq. (9) by expanding the root, using (7b), the approximate values of $K$, and regrouping.

The first application of Eq. (9) was to the $K = \frac{1}{2}-$ and $K = \frac{3}{2}-$ bands of $W^{183}$, which successfully explained the discrepancies between the experimental level spacings and the values given by neglecting the interband coupling [114]. This analysis has been extended to other odd-A nuclei; however, in each case, the effect of deformation vibrations has been ignored. This, then, is as far as we can proceed in the analysis of the level structure of these nuclei without solving the particle problem, the solution of which will be discussed in the following section.

## B. The Single-Particle Solutions

The problem remaining to be solved is that of the extra nucleon moving in the potential well of the deformed, but symmetric, core whose Schrödinger equation is

$$H_p\,|\,\Omega\rangle = E_\Omega\,|\,\Omega\rangle. \tag{11}$$

The complete particle Hamiltonian is defined in Eq. (3). The solutions can be determined as soon as the form of the potential $V(\mathbf{r}, \mathbf{l}, \mathbf{s})$ is given; however, their validity will depend upon how well the conditions assumed for separation of the original Hamiltonian are met. It is usual to accept this separation which depends upon the condition that the particle spacings are far greater than the rotational spacings and the interaction remaining, $H_c$ can be adequately accounted for by a perturbation calculation. Although a recipe such as this has been quite successful in assigning nuclear properties to odd-$A$ muclei in the deformed regions, it would be much more consistent to diagonalize the complete Hamiltonian. With high speed computers this is entirely possible but no such program has been carried out for such symmetric nuclei. A calculation such as this has been done for a restricted range of light nuclei as will be seen in the next section. Although that investigation did include the special case of axial symmetry, its extension to heavier nuclei has not been attempted.

We should expect that the form of the potential function in Eq. (3) should go over in a continuous fashion to the shell-model potential when the deformation goes to zero. That is, for the spherical case, the potential function should be that of a squared-off harmonic oscillator potential with a spin-orbit term. Also in the spherical limit we should expect the eigenvalue solutions $E_\Omega$ to reproduce the well-known magic numbers at nucleon number 2, 8, 20, 50, 82, 126, .... This last condition in itself is an important restriction on the form of the potential. The most often used potential is

$$V(\mathbf{r},\ \mathbf{l},\ \mathbf{s})_{\text{sym}} = m/2[\omega^2(x_1^2 + x_2^2) + \omega_3^2 x_3^2] + C\,\mathbf{l}\cdot\mathbf{s} + D\,\mathbf{l}^2 \quad (12)$$

from Nilsson [134] who, however, truncated his particle Hamiltonian by dropping the $\mathbf{j}\cdot\mathbf{j}$ term from Eq. (3). Axial symmetry is assured by setting $\omega_1 = \omega_2 = \omega$, which with $\omega_3$ is parameterized by

$$\begin{aligned} \omega^2 &= \omega_0^2(1 + 2/3\ \delta) \\ \omega_3^2 &= \omega_0^2(1 - 4/3\ \delta)\,, \end{aligned} \quad (13)$$

where $\delta$ is a measure of the deformation. The constant volume condition has been invoked since $\omega^2 \cdot \omega_3$ is proportional to the volume and this product is constant to first order. Indeed for an isotropic oscillator we can calculate the root mean square radius in closed form and relate it to the nuclear radius $R_0 = 1.2\ A^{1/3} \times 10^{-13}$ cm. The oscillator strength is then

$$\hbar\,\omega_0 = \eta\ A^{-1/3}\ \text{MeV}.$$

The constant is often taken as $\eta = 41$; however, in actual fact it is to be determined once the theoretical spectrum is compared with the experimental one as $\hbar\omega_0$ is, from the point of view of theories such as these, but a scale factor.

The oscillator deformation parameter $\delta$ can be related to the core deformation parameter $\beta$ only by assuming some definite relation between nuclear surface and oscillator potential. One such assumption takes the nuclear surface to be an equipotential from which, to first order

$$\delta = \frac{3}{2}\sqrt{\frac{5}{4\pi}}\,\beta \cong 0.946\,\beta\,.$$

(It is important to note that in the literature of this subject different authors use different deformation parameters and, as is to be expected, some authors use more than one. A table relating some of these various conventions is to be found in Mottelson and Nilsson [34], which of course is not all inclusive.)

The Hamiltonian to be used can now be written as

$$H = H_0 + C\mathbf{l}\cdot\mathbf{s} + D\mathbf{l}^2\,, \tag{14}$$

where the anisotropic oscillator Hamiltonian $H_0$ is defined by

$$H_0 = \frac{p^2}{2m} + \frac{m}{2}\left[\omega^2(x_1^2 + x_2^2) + \omega_3^2 x_3^2\right]. \tag{15}$$

Dimensionless coordinates are conveniently introduced at this point

$$\xi_k = \sqrt{\frac{m\omega_0}{\hbar}}\,x_k\,, \qquad k = 1, 2, 3$$

$$\varrho^2 = \sum_{k=1}^{3} \xi_k^2$$

so that $H_0$ becomes

$$\begin{aligned} H_0 &= H_{00} + H_\beta \\ H_{00} &= -\,(\hbar\omega_0/2)\,(\nabla^2 - \varrho^2) \\ H_\beta &= -\,\hbar\omega_0\beta\varrho^2 Y_{20}(\theta, \varphi). \end{aligned}$$

It is most convenient to carry out the diagonalization of the Hamiltonian (14) in a basis in which $H_{00}$ is itself diagonal. There are two such bases. One is analogous to the cartesian basis of the familiar three-dimensional oscillator problem; the other is similar to the spherical polar basis. Both are used. The first basis is associated with the three commuting operators

$$H_{00_k} = -\frac{1}{2}\left(\frac{d^2}{d\xi_k^{\,2}} - \xi_k^{\,2}\right), \qquad k = 1, 2, 3 \tag{16a}$$

$$\sum_{k=1}^{3} H_{00_k} \mid n_1, n_2, n_3\rangle = E_{00} \mid n_1, n_2, n_3\rangle \tag{16b}$$

where

$$E_{00} = \hbar\omega_0\left(N + \frac{3}{2}\right) = \hbar\omega_0\left(\sum_{k=1}^{3} n_k + \frac{3}{2}\right). \tag{16c}$$

We shall return to this basis later. The alternative basis, with respect to which the diagonalization has been carried out, simultaneously diagonalizes the operators $H_{00}$, $\mathbf{l}^2$, $l_3$, and $s_3$ whose eigenvalues are

$$\begin{aligned} H_{00} \mid Nl\Lambda\Sigma\rangle &= \hbar\omega_0(N + \tfrac{3}{2}) \mid Nl\Lambda\Sigma\rangle \\ \mathbf{l}^2 \mid Nl\Lambda\Sigma\rangle &= l(l+1) \mid Nl\Lambda\Sigma\rangle \\ l_3 \mid Nl\Lambda\Sigma\rangle &= \Lambda \mid Nl\Lambda\Sigma\rangle \\ s_3 \mid Nl\Lambda\Sigma\rangle &= \Sigma \mid Nl\Lambda\Sigma\rangle. \end{aligned} \tag{17}$$

As usual the quantum numbers $N$ and $l$ are restricted to the ranges

$$\begin{aligned} N &= 0,\ 1,\ 2,\ \ldots \\ l &= N, N-2, N-4, \ldots, \qquad 1 \quad \text{or} \quad 0 \end{aligned}$$

and the condition of axial symmetry requires that

$$\Omega = \Lambda + \Sigma.$$

Two new parameters are defined in terms of those of Eq. (14)

$$\begin{aligned} \varkappa &= C/\hbar\omega_0 \\ \mu &= D/\varkappa\hbar\omega_0. \end{aligned}$$

It is then necessary, in this basis, to diagonalize only the operator

$$H_\beta + C\mathbf{l}\cdot\mathbf{s} + D\mathbf{l}^2 = H' = -\hbar\omega_0[\beta\varrho^2 Y_{20}(\theta, \varphi) + 2\varkappa\mathbf{l}\cdot\mathbf{s} + \mu\varkappa\mathbf{l}^2]. \tag{18}$$

From the form of the potential it is clear that parity, as well as the quantum number $\Omega$, are conserved quantities so that matrix elements with $\Delta N = 1$ vanish but those with $\Delta N = 2$ may not. In the approximation of Nilsson these latter are consistently set equal to zero.

The parameters $\varkappa$ and $\mu$ are selected in such a way that for zero deformation, the shell model ordering of states is obtained. In order to achieve this Nilsson took a single value of $\varkappa = 0.05$ and different values of $\mu_N$ for each shell. It has been found that for light nuclei (in particular those in the $2s - 1d$ shell), $\varkappa = 0.08$ is a better value than the original one.

In Appendix D are tables giving the normalized eigenvector coefficients $C_{j\Omega=K}$ defined by

$$\begin{aligned} H \mid \Omega\rangle &= E_\Omega \mid \Omega\rangle \\ &= E_\Omega \sum_j C_{j\Omega} \mid Nlj\Omega = K\rangle. \end{aligned}$$

The $C_{j\Omega}$ are related to the normalized $a_{l\Lambda}$ of Nilsson by the usual vector coupling relation between coupled and uncoupled representations. Figures D-1–D-5 give the eigenvalues for deformations on the range $-0.4 \leq \beta \leq 0.4$. For $N \leq 2$ the value $\varkappa = 0.08$ has been used whereas for $N > 2$ the value $\varkappa = 0.05$ has been used. Furthermore, each value of $\mu_N$ is given in each table. The eigenvalues $E_\Omega$ have been plotted as a function of $\beta$ and appear facing each table. Associated with each level are a much-used set of numbers, the asymptotic quantum numbers $N, n_z, \Lambda, \Omega$, and the parity. Usually this set of numbers is used with the notation $\Omega, \pi\ [N, n_z, \Lambda]$, which will be followed here although the parity is redundant since it is positive or negative as $N$ is even or odd, respectively. These quantum numbers label the particle levels for very large deformations, for then the $\mathbf{l} \cdot \mathbf{s}$ and $\mathbf{l}^2$ terms may be treated as perturbations; the levels in lowest order are simply those of a pure anisotropic oscillator. The asymptotic quantum numbers are easily assigned in each oscillator shell by observing that for large positive deformations the lowest level for which $\Omega = \frac{1}{2}$ has $n_z = N$, the next lowest level has $n_z = N - 1$. The process continues in this way until the highest level which has $n_z = 0$. The next largest value $\Omega = \frac{3}{2}$ has for the lowest particle level $n_z = N - 1$, proceeding until the highest level has $n_z = 0$. We can therefore assign at once all of the $n_z$ values from which the $\Lambda$ values follow, since $\Lambda$ is even or odd as $(N - n_z)$. (Recall that in every case

$\Omega = \Lambda + \Sigma$ is always a good quantum number.) Therefore the lowest $\Omega = \frac{1}{2}$ level is to be labeled $\frac{1}{2}(-1)^N[N, N, 0]$, while the lowest $\Omega = \frac{3}{2}$ level bears the label $\frac{3}{2}(-1)^N[N, N-1, 1]$.

The eigenvalues of the Hamiltonian (18) can be given in closed form for two cases. The first is associated with the asymptotic quantum numbers $\frac{1}{2}(2N+1)\,(-1)^N[N, 0, N]$ ($\Omega$ has its maximum value in the shell) with eigenvalue

$$E' = \varkappa\hbar\omega_0 N\left[\frac{\beta}{2\varkappa}\sqrt{\frac{5}{4\pi}} - \mu_N(N+1) - 1\right].$$

For the next lowest value of $\Omega$ the eigenvalues are

$$E' = \frac{1}{2}\varkappa\hbar\omega_0\left[\frac{\beta}{2\varkappa}\sqrt{\frac{5}{4\pi}}(2N-3) - 2\mu_N(N+1) + 1 \pm\sqrt{\left(\frac{3\beta}{2\varkappa}\sqrt{\frac{5}{4\pi}} + 2N - 1\right)^2 + 8N}\,\right]$$

the asymptotic quantum numbers being $(2N-1)/2(-1)^N\,[N, 0, N]$ for the plus sign and $\frac{1}{2}(2N-1)\,(-1)^N\,[N, 1, N-1]$ for the minus sign. The other eigenvalues are best obtained by the diagonalization of the appropriate matrix. (Tables in the literature often use the parameter $\eta = (3\beta/2\varkappa)\sqrt{5/4\pi}$ instead of $\beta$.) In each case, to these eigenvalues we must add $\hbar\omega_0(N+\frac{3}{2})$ to get the total energy of the states. In Appendix D the eigenvalues are given in units of $\hbar\omega_0$ since this parameter is different for each nucleus.

The Hamiltonian (14) differs from the Hamiltonian $H_p$ of Eq. (3) in that the $\mathbf{j}\cdot\mathbf{j}$ term is omitted in the former. That this omission may rearrange these intrinsic particle levels can be seen by writing $H_p$ as

$$H_p = H + \frac{\hbar^2}{2\mathscr{I}_0}\,\mathbf{j}\cdot\mathbf{j} = H + \frac{\hbar^2}{2\mathscr{I}_0}\,(\mathbf{l}^2 + 2\mathbf{l}\cdot\mathbf{s} + \mathbf{s}^2).$$

Since $\mathbf{s}^2$ is diagonal in this representation with eigenvalue 3/4 this expression may be written

$$\frac{H_p}{\hbar\omega_0} = N + \tfrac{3}{2} + \tfrac{3}{4}\,\varepsilon - \beta\varrho^2 Y_{20}(\theta, \varphi) - 2(\varkappa - \varepsilon)\,\mathbf{l}\cdot\mathbf{s} - (\mu\varkappa - \varepsilon)\mathbf{l}^2 \tag{19}$$

where $\varepsilon = \hbar/2\mathscr{I}_0\omega_0$. For nuclei in the $2s - 1d$ shell taking $A = 25$, we find typical values of $\hbar^2/2\mathscr{I}_0$ of about 300 keV so that $\varepsilon \cong 2 \times 10^{-2}$ and is of the same order of magnitude as $\varkappa$ and $\varkappa\mu$. The omission of the $\mathbf{j} \cdot \mathbf{j}$ term has been found, by direct diagonalization of the complete Hamiltonian, to cause a reordering and shifting of the particle levels in this shell. [62] For higher shells it will be less significant. From the example $Tb^{159}$, $N = 4$, $\mu_4 = 0.55$, thus $\mu\varkappa = 2.75 \times 10^{-2}$ but $\varepsilon$ is $1.5 \times 10^{-3}$ more than an order of magnitude less than either $\varkappa$ or $\mu\varkappa$. Therefore, for the rare earth and actinide deformed regions the omission of the $\mathbf{j} \cdot \mathbf{j}$ term in the particle Hamiltonian is reasonable in that the lower intrinsic levels are little affected. For the $2s - 1d$ shell, dropping this term is probably not wise, although, as will be seen, the simple Nilsson model is not adequate and we must resort to more complicated calculations.

In order to show how the Nilsson levels can be used to predict properties of odd-$A$ nuclei we consider two examples: ${}_{31}Ga^{69}_{38}$ and ${}_{74}W^{183}_{109}$. The former is in a region known to have little or no deformation, whereas the latter is quite deformed. Since $| \Omega |$ is a good quantum number, the Pauli principle permits two like particles in each intrinsic level. Thus counting up from the bottom of the Nilsson diagram by twos, it is found that the 31st proton can be put into an $\Omega = \frac{3}{2} -$ level (the known ground state spin and parity) only for $\beta$ very small and positive ($0 < \beta \lesssim 0.05$) or for $\beta$ very large and negative ($\beta \gtrsim - 0.3$). The first assignment is the proper one and is in keeping with the fact that this region is generally considered to be one of spherical, or vibrational, nuclei.

For the other example, $W^{183}$ has a ground state spin and parity of $\frac{1}{2} -$ which, as we have seen, forms the basis of a $K = \frac{1}{2}$ band. It is found by counting up that the 109th neutron will enter such a level if the deformation is about 0.2. Examining the appropriate level diagram in Appendix D shows that four intrinsic levels are very close together. They are the $\frac{1}{2} -$ [510], $\frac{3}{2} -$ [512], $7/2 -$ [503], and the $\frac{1}{2} +$ [651] levels. We should expect that in the level structure evidence for each of the particle levels will appear as the head of an appropriate $K$ band. This is in fact the case with only the $\frac{1}{2} +$ level missing; however, the $9/2 +$ [634] level appears instead. The assignments are as follows: ground state $\frac{1}{2} -$ [510], 208.8 keV state $\frac{3}{2} -$ [512], 453.1 keV state $7/2 -$ [503], 309.5 keV state $9/2 +$ [634]. In order to achieve this last assignment it is necessary to break a pair of neutrons in the core, exciting one to the $\frac{1}{2} -$ [510] level which in turn leaves the 107th neutron unpaired in the

9/2 + [634] orbit to furnish the level characteristics. Upon each of these intrinsic levels is to be built a rotational band; this accounts for the remaining low-lying levels in this nucleus.

With the particle problem solved, the state function components $C_{j\Omega}$ are known and the various mixing parameters can be calculated. For the case of $W^{183}$, the decoupling parameter $a$ of Eq. (8a) can be calculated since the particle orbitals and the deformation have been determined. Nilsson has calculated the range of this parameter for his original solutions of the particle problem [134] and the empirical value determined for the $\frac{1}{2}-$ [510] level of 0.19 falls within the stated range. From Eq. (9a) the band mixing parameter $A_K$ can also be gotten. Similar calculations can be made for other deformed nuclei making use of the tables of Appendix D.

Not only can we find rotational levels built upon these intrinsic states in odd-$A$ nuclei but we should expect to find three other bands built upon a given particle state with $\Omega = K_0$. One will be a beta vibrational band with $K = K_0$ whereas the other two could be called gamma vibrational bands with $K = K_0 \pm 2$; the parity in each case should be the same as the intrinsic level. Although no detailed calculations have been published concerning the nature of such excited levels, they have been observed in both the rare earth and actinide deformed regions and it is possible to make some qualitative statements concerning these states. The beta band ($K = K_0$) arises from the coupling of the intrinsic level to the first excited beta band in the core with the band head 012 level. The $K = K_0 \pm 2$ bands will be based on the 221 level of the core. Furthermore, one should expect that those levels, because of their origin, will be mainly collective in nature so that the $E2$ transition moments to and from them will be considerably higher than to nearby intrinsic levels. That is, they should be easier to coulomb excite. Also the levels in the beta band should be considerably easier to excite than those in the gamma bands, since this is generally true of the neighboring even-even nuclei.

An example of such vibrational bands is to be found in ${}_{67}\mathrm{Ho}^{165}_{98}$ with the ground state characterized by the orbital $\frac{7}{2}-$ [523] (associated with which five excited rotational levels are known). Both of the gamma vibrational bands appear with the $K = K_0 - 2 = \frac{3}{2}-$ band head at 514 keV and with the $K = K_0 + 2 = 11/2-$ band head at 687 keV. Both bands have been coulomb excited [82] and are certainly associated with the ground state band as their inertial parameters ($\hbar^2/2\mathscr{I}_0$) are 10.4 keV and 10.2 keV. For the ground state band the value is 10.5 keV. Further evidence of this relationship is found from the transition moments. It is curious that the analog to the beta band in even nuclei has not been observed. Indeed, almost no beta bands are known in odd-$A$ nuclei.

The shape of the nuclear energy surface can be determined, at least near equilibrium, by using particle eigenvalues to calculate the total nuclear energy $\mathscr{E}$. Since $\mathscr{E}$ is the sum of the kinetic and potential energies for all $A$ particles, it will be a function of the deformation. The equilibrium deformation $\beta_0$ will be determined by setting $\partial\mathscr{E}(\beta)/\partial\beta = 0$. In order to find the ground state equilibrium deformation, all particle orbitals below the Fermi surface are assumed filled. However, by leaving vacant orbitals near the top of this surface, that is, by promoting particles to unfilled states, it is possible to see how the deformation will change with particle excitation.

Confining the discussion to two-particle forces

$$V_i = \sum_{i \neq j}^{A} V_{ij}$$

the total energy function can be written

$$\begin{aligned} \mathscr{E}(\beta) &= \sum_{i=1}^{A} T_i + \frac{1}{2} \sum_{i \neq j}^{A} V_{ij} \\ &= \frac{3}{4} \sum_i \mathscr{E}_i(\beta) - \frac{1}{4} \sum_i (\bar{V}_i - \bar{T}_i). \end{aligned} \tag{20}$$

The $\mathscr{E}_i(\beta)$ are the particle eigenvalues so that it is only necessary to evaluate the second sum in (20). In order to do this it is necessary to remember that the original diagonalization neglected the coupling between shells with $\Delta N = 2$. For heavy nuclei Figs. D-4 and D-5 show that there is a great deal of intershell coupling (evidenced by levels with $\Delta N = 2$ crossing). The major part of this coupling between shells can be accounted for by a change in the representation. Let

$$\eta_k = \sqrt{\frac{m\omega_k}{\hbar}}\, x_k\,, \qquad k = 1, 2, 3$$

$$\sigma^2 = \sum_{k=1}^{3} \eta_k{}^2$$

and define a new deformation parameter $\varepsilon$ by

$$\omega_1 = \omega_2 = \bar{\omega}_0 \left(1 + \frac{1}{3} \varepsilon\right)$$
$$\omega_3 = \bar{\omega}_0 \left(1 - \frac{2}{3} \varepsilon\right),$$

which is equal to $\delta$ in first order. The Hamiltonian (15) can be written in a cartesian-like basis in which it separates. This basis is superficially similar to that of Eqs. (16) and defines the state vectors $| n_1', n_2', n_3' \rangle$ and a new principal quantum number $N'$, where as usual

$$N' = \sum_{i=1}^{3} n_i'.$$

The advantage gained is that now $H_0$ is diagonal in $N'$ whereas before only $H_{00}$ was diagonal in $N$. To see this, let

$$\nabla_\eta^2 = \sum_{k=1}^{3} \frac{\partial^2}{\partial \eta_k^2}$$
$$H_0 = \frac{1}{2} \hbar\bar{\omega}_0 \left[\left(1 + \frac{\varepsilon}{3}\right)(- \nabla_\eta^2 + \sigma^2) + \varepsilon\left(- \frac{\partial^2}{\partial \eta_3^2} + \eta_3^2\right)\right]. \quad (21)$$

The first term is obviously diagonal in $N'$ whereas the second term is diagonal in $n_3'$ since both are simple three- and one-dimensional oscillators. Since the second term is also diagonal in $n_1'$ and $n_2'$ then it is diagonal in $N'$.

By introducing angular momentum-like operators

$$l_j' = - i\left(\eta_k \frac{\partial}{\partial \eta_l} - \eta_l \frac{\partial}{\partial \eta_k}\right)$$

we may rewrite $H_0$ in a spherical-like coordinate system and use for the base vectors $| N', l', \Lambda', \Sigma \rangle$. The Hamiltonian (14) is then

$$\begin{aligned} H = {} & H_0 + C\mathbf{l}' \cdot \mathbf{s} + D\mathbf{l}'^2 \\ & + C(\mathbf{l} - \mathbf{l}') \cdot \mathbf{s} + D(\mathbf{l}^2 - \mathbf{l}'^2) = H' + H_1. \end{aligned} \quad (22)$$

Clearly $H'$ has the same matrix elements in the $| N, l', \Lambda', \Sigma \rangle$ basis as $H$ does in the $| N, l, \Lambda, \Sigma \rangle$ basis so that for the energy surface calculation the tabulated eigenvalues can still be used. Nilsson has shown that the effect of the perturbation Hamiltonian $H_1$ in intershell coupling is only about 10% of $H_\beta$ so that this new representation does take into account most of the interaction between shells with $\Delta N = 2$.

In this basis we now split the second term of Eq. (20) into angular momentum dependent and independent terms

$$\sum_i (\bar{V}_i - \bar{T}_i) = \sum_i (\bar{V}_i' - \bar{T}_i') + C\mathbf{l}' \cdot \mathbf{s} + D\mathbf{l}'^2,$$

where the primes stand for the new basis. Since for an oscillator the average values of the potential minus kinetic energies is zero, Eq. (20) becomes just

$$\mathscr{E}(\beta) = \frac{3}{4} \sum_i \mathscr{E}_i(\beta) - \frac{1}{4} \langle C\mathbf{l}' \cdot \mathbf{s} + D\mathbf{l}'^2 \rangle. \tag{23}$$

The $\mathscr{E}_i(\beta)$ are tabulated and the diagonal matrix elements from the second term are for large deformations

$$\langle n_1' + n_2'n_3'\Lambda\Sigma \mid \mathbf{l}' \cdot \mathbf{s} \mid n_1' + n_2'n_3'\Lambda\Sigma \rangle = \Lambda\Sigma$$
$$\langle n_1' + n_2'n_3'\Lambda\Sigma \mid \mathbf{l}'^2 \mid n_1' + n_2'n_3'\Lambda\Sigma \rangle$$
$$= \Lambda^2 + 2(n_1' + n_2')n_3' + n_3' \mid N'.$$

The principal effects neglected in Eq. (23) are residual interactions such as pairing forces and the coulomb energy between protons. In Chapter 1 it was seen that the change in coulomb energy is quadratic in the deformation parameter

$$\Delta E_c = - A\beta^2, \qquad A > 0.$$

Thus the inclusion of this term will increase the equilibrium deformation. However, the bulk of these effects are due to the core and have been taken into account in a rough way by the fact that different values of $\mu_N$ have been used for different shells and in the higher shells different values of $\mu_N$ are used for protons and neutrons. Very extensive equilibrium shape studies have been made by Mottelson and Nilsson [34] and the validity of the various approximations and the model in general is borne out by the very good agreement between their calculations and experiment. These calculations not only reproduce the general trend of deformation with $A$ showing the observed preference for prolate equilibrium shapes in the deformed regions, they also yield many of the ground state properties as well as many of the gamma and beta branching ratios, as will be seen later.

A point of somewhat more than historical interest is the fairly large gap in the intrinsic level diagram around 152 neutrons and positive deformations (Fig. D-5). This particle number occurs, in spherical nuclei, in the region between the $g_{9/2}$ and $j_{15/2}$ shells; however, there is no "magic number" for spherical nuclei in this region although a gap is to be associated with a shell closing. In the deformed nuclei in this region of the actinides this gap is to be associated with a subshell closing. Historically, the existence of this gap was suggested by Ghiorso and his co-workers [102] about a year before the first publication of the intrinsic particle levels in deformed systems. The original suggestion was based upon alpha-decay energies (see Chapter 7) but further support for such a gap is based upon a sudden decrease in thermal neutron capture cross sections in the californium isotopes at $N = 152$ as well as a change in the spontaneous fission half-life there. These are all signs of shell closings.

Even though this model of a nucleon coupled to a symmetrically deformed core has been used very successfully in the rare earth and actinide deformed regions, it has been used considerably less successfully in light nuclei. Indeed this region, in particular the $2s - 1d$ shell, was the last region to which the model was applied. The initial proposal was made by Litherland and co-workers [125] as the result of an extensive study of $Al^{25}$. Subsequently, many of the other nuclei in this shell were studied from this point of view. At least below $S^{32}$ there seems little doubt that these nuclei are deformed and show some sort of rotational structure. Indeed, this strong coupling model can be made to explain the quantum number assignments to ground and low-lying states. The major difficulty in applying such simple models as these is that the wide spacing of levels for light nuclei leads to a strong overlap of various $K$ bands. In most of the nuclei in this shell, $\Delta K = 1$ for these bands so that the separation of core and intrinsic particle parts of the problem by neglecting $H_c$ is not valid. It is possible to carry out band-mixing calculations in this region using the methods discussed in Section A. However, the results depend to a great extent upon the number of bands mixed. This mixing technique will not do for the mirror pair at $A = 25$, for here the known positive parity $K$ bands have $\Delta K = 0,2$ and the operator $H_c$ will not mix them. Thus the calculations are somewhat simplified.

An important feature of these calculations in this shell is the fact that the observed negative parity levels are associated with intrinsic levels from the $2p - 1f$ shell that descend into the $2s - 1d$ shell for sufficiently large deformations. In other calculations (Section C) this method does not work, as the nearest $N = 3$ levels are much too far away. In Section C it will be seen that this more complicated model is very successful in dealing with the positive parity levels of deformed light nuclei.

## C. Models Without Axial Symmetry

If the properties of some even-even nuclei are best explained by the asymmetric collective model, then there is no *a priori* reason to exclude core deformations like these when considering neighboring odd-$A$ nuclei. However, the analytic complexities associated with such a system's Schrödinger equation increase greatly. The simple and useful separation of the symmetric core Hamiltonian into Eqs. (2)–(4) arises from the requirement that $K = \Omega$. The total collective Hamiltonian is still a sum of rotational and particle parts

$$H_R = \frac{\hbar^2}{2} \sum_{k=1}^{3} L_k^2/\mathscr{I}_k = \frac{\hbar^2}{2} \sum_{k=1}^{3} (I_k - j_k)^2/\mathscr{I}_k \tag{24}$$

$$H_p = \frac{p^2}{2m} + V(\mathbf{r}, \mathbf{l}, \mathbf{s}). \tag{25}$$

The potential function in the particle Hamiltonian (25) can be the same as for the symmetric case except that the condition of axial symmetry is relaxed.

$$V(\mathbf{r}, \mathbf{l}, \mathbf{s}) = \frac{m}{2} \sum_{i=1}^{3} \omega_i^2 x_i^2 + C\mathbf{l} \cdot \mathbf{s} + D\mathbf{l}^2, \tag{26}$$

By abandoning axial symmetry the simple separation of the Hamiltonian is no longer valid since the rotational Hamiltonian will have a strong mixing effect on particle states. However, if these states are widely separated so that we can assume little mixing by $H_R$ then it is useful to inquire whether or not the energy surface generated by $\mathscr{E}(\beta, \gamma)$ has minima for values of $\gamma$ not zero. A program such as this is similar to that outlined before to find the equilibrium ground state deformation except that $\mathscr{E}(\beta)$ of Eq. (23) is replaced by $\mathscr{E}(\beta, \gamma)$. Again by making the nuclear surface an equipotential, the oscillator frequencies to first order in $\beta$ are

$$\omega_k = \omega_0\left[1 - \sqrt{\frac{5}{4\pi}}\,\beta\cos(\gamma - 2\pi k/3)\right], \quad k = 1, 2, 3 \tag{27}$$

instead of Eq. (13). Thus for the nondiagonal part of the Hamiltonian of Eq. (18), the $-\beta\varrho^2 Y_{20}(\theta, \varphi)$ term is replaced by

$$-\beta\varrho^2\left[\cos\gamma\, Y_{20}(\theta, \varphi) + \frac{\sin\gamma}{\sqrt{2}}(Y_{22}(\theta, \varphi) + Y_{2-2}(\theta, \varphi))\right]. \tag{28}$$

Therefore the resulting diagonalization now depends on the new parameter $\gamma$ as well as on $\beta$, $\varkappa$, and $\mu_N$. The results of an investigation of asymmetric single-particle wells has been reported by Newton [133]. He found that the equilibrium shape for such "model" nuclei near $Z = N = 12$ and 20 were asymmetric ($\gamma \neq 0$) and probably asymmetric in the region $A = 44$–$46$. Instead of using Nilsson's values of $\varkappa$ and $\mu_N$, Newton varied $\varkappa$ (by varying $C = -2\varkappa\hbar\omega_0$ but retaining Nilsson's value of $\mu_N \propto D$) the results being displayed in a number of graphs. Since the potential function is no longer axially symmetric, $\Omega$ is not a good quantum number. Thus the single-particle states must now be represented by

$$\begin{aligned} |\,\nu\rangle &= \sum_{\Omega} d_\Omega\,|\,\Omega\rangle \\ &= \sum_{j,\Omega} C'_{j\Omega}\,|\,j\Omega\rangle. \end{aligned} \tag{29}$$

The next approximation of interest would be to add the effect of the rotational Hamiltonian (24). If the assumption that $H_R$ produces little mixing of particle states is valid then the state functions will be labeled by the total spin, energy, and particle state and are of the form

$$|\,IME\nu\rangle = \sum_K{}' A_K\{D^I_{MK}\,|\,\nu\rangle + (-1)^{I-j}\,D^I_{M-K}\,|\,-\nu\rangle\} \tag{30}$$

where the prime on the sum means that $K$ is restricted to values such that $K - \Omega = 2n$, $n$ integral. In effect, we just make use of Newton's solutions with the solution of the core problem. An extensive examination of this approximation has been carried out by Hecht and Satchler with parameters valid for the rare earth deformed region [106]. In particular they have applied their rather extensive calculations to the nucleus $Pt^{195}$

which has a ground state spin $\frac{1}{2}$ —; the model fits the level sequence moderately well. Other applications to nuclei in the region $150 \leq A \leq 190$ have been reported; however, this model has not been extended into the actinide region. Both the calculations of Newton and of Hecht and Satchler made use of the oscillator basis of Eq. (21) and (22) which takes into account the interaction between different oscillator shells.

The most consistent collective model for these odd-$A$ nuclei would be one that makes no assumption about the relative separation of particle and rotational levels but instead diagonalizes the complete Hamiltonian. The logical way to proceed is to assume first only that the even-even core is rigid against deformation and asymmetry vibrations and then relax this assumption of rigidity. The sum of the core and particle Hamiltonians (24) and (25) with the potential function (26) is just

$$\begin{aligned} H = (-\nabla^2 + \varrho^2) + \hbar\omega_0 \Big\{ & \frac{\hbar^2}{8B_2\beta^2\hbar\omega_0} \sum_{k=1}^{3} \frac{(I_k - j_k)^2}{\sin^2(\gamma - 2\pi k/3)} \\ & - \beta\varrho^2 \left[ \cos\gamma Y_{20}(\theta, \varphi) + \frac{\sin\gamma}{\sqrt{2}} (Y_{22}(\theta, \varphi) + Y_{2-2}(\theta, \varphi)) \right] \\ & - 2\varkappa\, \mathbf{l} \cdot \mathbf{s} - \mu\varkappa \mathbf{l}^2 \Big\} \end{aligned} \tag{31}$$

$\varrho$ having been defined before in connection with the basis (17). On introducing a core mass parameter $P$ by

$$P = \hbar^2/8B_2\beta^2\varkappa\hbar\omega_0 \tag{32}$$

the nondiagonal part of (31) $H_{\beta\gamma}$ is just

$$H_{\beta\gamma} = \hbar\omega_0(h_R + h_P) \tag{33}$$

with

$$h_R = \varkappa P \sum_{k=1}^{3} (I_k - j_k)^2/\sin^2(\gamma - 2\pi k/3) \tag{34a}$$

$$\begin{aligned} h_P = & - \beta\varrho^2 \left[ \cos\gamma Y_{20}(\theta, \varphi) + \frac{\sin\gamma}{\sqrt{2}} (Y_{22}(\theta, \varphi) + Y_{2-2}(\theta, \varphi)) \right] \\ & - 2\varkappa\mathbf{l} \cdot \mathbf{s} - \mu\varkappa\mathbf{l}^2. \end{aligned} \tag{34b}$$

The reason for the choice of mass parameter (32) is now clear since the level structure is unaffected by changes in the parameter $\varkappa$. If $\varkappa$ is changed by some fraction to $\varkappa f$ then changing $\hbar\omega_0$ to $\hbar\omega_0 f$ and $\beta$ to $\beta/f$ multiplies the Hamiltonian (33) by $f$ and thus the eigenvalues by $f$, leaving the level sequence and relative spacings the same. The operator (33) is to be diagonalized in terms of the symmetrized angular momentum basis

$$| IMKj\Omega\rangle = (1/\sqrt{2})\,[D^I_{MK} \,|\, j\Omega\rangle + (-1)^{I-j} D^I_{M-K} \,|\, j - \Omega\rangle] \quad (35)$$

with the symmetry restrictions

$$K - \Omega = 2n, \qquad n = 0, 1, 2, \ldots$$
$$\Omega > 0$$

but

$$K = \Omega \quad \text{for} \quad \gamma = 0.$$

The Hamiltonian (33) has the property

$$H(\beta, \gamma) = H(-\beta, 60^\circ - \gamma).$$

Therefore we may take the parameter ranges to be

$$P \geq 0$$
$$\beta \geq 0$$
$$0 \leq \gamma \leq \pi/3.$$

Unfortunately the solution of the deformed-core problem does not follow immediately from the diagonalization of the Hamiltonian (33), since no account has been made of the Pauli principle. For the symmetric core problem, however, it is quite simple to let like nucleons enter each intrinsic level pairwise. The solution here depends upon removing elements from the basis (35) and the final results are apparently quite dependent upon the basis truncation procedure used. Thus adding two particles to a level fills it and from the point of view of these collective models the two nucleons enter the core. By removing appropriate basis elements, the probability of finding the outer nucleon in a filled level is identically zero. The truncation procedure depends upon identifying the particle quantum numbers of the just filled level at zero deformation. An example from the $N = 1$ oscillator shell should demonstrate the method. In Table 4-1 is displayed the complete basis for $I = \frac{3}{2}$ and $N = 1$ and reference should also be made to Fig. D-1. The odd-nucleon number $\zeta$ is defined to be $Z$ or $N$, whichever is odd. Then for the $N = 1$ oscillator shell $2 < \zeta < 8$ and for $He^5$ and all of the odd lithium isotopes

TABLE 4–1

The Complete Basis for $I = \frac{3}{2}$ in the $N = 1$ Oscillator Shell

| Basis number | $\vert I$ | $K$ | $j$ | $\Omega\rangle$ |
|---|---|---|---|---|
| 1 | 3/2 | 3/2 | 3/2 | 3/2 |
| 2 | 3/2 | −3/2 | 3/2 | 1/2 |
| 3 | 3/2 | −3/2 | 1/2 | 1/2 |
| 4 | 3/2 | −1/2 | 3/2 | 3/2 |
| 5 | 3/2 | 1/2 | 3/2 | 1/2 |
| 6 | 3/2 | 1/2 | 1/2 | 1/2 |

$\zeta = 3$ that is one nucleon outside the $He^4$ closed shell. For these nuclei the entire basis of Table 4-1 is used. For $Be^9$ and the boron isotopes $\zeta = 5$, three particles outside of the closed shell. Two of these fill the $\frac{1}{2} -$ [110] level (assuming positive deformation, otherwise the $\frac{3}{2} -$ [101] level is filled and the following argument is similar) making it unavailable for more particles. At zero deformation this level has the character $j = \frac{3}{2}$, $\Omega = \frac{1}{2}$, and all basis elements are removed with these quantum numbers. That is, elements numbered 2 and 5 in Table 4-1 are removed. For $C^{13}$ and the nitrogen isotopes $\zeta = 7$, there are five nucleons outside the closed shell of which four fill levels $\frac{1}{2} -$ [110] and $\frac{3}{2} -$ [101]. These levels have character $j = \frac{3}{2}$, $\Omega = \frac{1}{2}$, and $\Omega = \frac{3}{2}$, respectively, at zero deformation so that basis elements 1, 2, 4, and 5 are removed. The properties of the nuclei in question are determined by the interaction of the core with basis elements 3 and 6.

For the untruncated basis the matrices to be diagonalized are of order $(N + 1)(N + 2)(2I + 1)/4$ so that machine diagonalization is necessary. For a given $I$ this diagonalization traces out a hypersurface in $P$, $\beta$, $\gamma$, $E$ space. Calculations and applications have only been reported for nuclei in the $2s - 1d$ shell and for the lowest $N = 3$ levels [62]. The results seem to reproduce quantitatively the observed level structure. The "goodness" of the model can be assessed from the fact that not any arbitrary spin sequence could be reproduced (that is, given some spin sequence $I_i$, $I$, ..., $I_n$, there might be no values of $P$, $\beta$, and $\gamma$ which would yield this set of values). Indeed the regions in $P$, $\beta$, $\gamma$, $E$ space where a fit to real nuclei was obtained were in general small. Whether or not the model would be equally successful in higher oscillator shells is moot.

CHAPTER 5

# Models of Odd-Odd Nuclei

The extension of the ideas of the previous chapter to deformed odd-odd-$A$ nuclei is simple and straightforward. Indeed we shall make essentially the same assumptions about strong coupling, axial symmetry, nature of the even-even core, etc. However, an additional assumption about the interaction between the two odd particles, the neutron and proton, is necessary and we assume that each moves independently of the other in the same deformed potential well. The correctness of this assumption will rest upon how well the model predicts experimentally measurable quantities. As we shall see in this chapter the model is quite accurate in its predictions of the energy level structure and, as we shall see in the next chapter, quite accurate in its predictions of the various measured, static, electromagnetic moments.

Since the development of the theory will follow that of the previous chapter, we may defer until later the form of the deformed potential. However, because of the success of the Nilsson potential for odd-$A$ nuclei, we shall in due course utilize it. This procedure has the advantage of displaying those properties that are inherent in the model and hence independent of the form of the potential.

Again $\mathbf{I}$ is the total angular momentum, $\mathbf{L}$ is the angular momentum of the even-even core and $\mathbf{j}_p$ and $\mathbf{j}_n$ are the angular momenta of the odd proton and neutron. None of these last three are constants of the motion but $\mathbf{I}$, which is their sum, is

$$\mathbf{I} = \mathbf{L} + \mathbf{j}_p + \mathbf{j}_n. \tag{1}$$

The requirement of strong coupling along with the independence of neutron and proton motion implies that

$$[I_1, I_2] = -iI_3, \text{ cyclically}$$
$$[j_{p_1}, j_{p_2}] = ij_{p_3}, \text{ cyclically}$$
$$[j_{n_1}, j_{n_2}] = ij_{n_3}, \text{ cyclically}$$
$$[\mathbf{j}_p, \mathbf{j}_n] = 0.$$

Again we take $M$ to be the projection of $\mathbf{I}$ upon the laboratory $Z$ axis, and $K$ its projection on the body-fixed 3 axis; $\Omega_p$ and $\Omega_n$ are the projections of $\mathbf{j}_p$ and $\mathbf{j}_n$ on this latter axis. Since we assume axial symmetry, $\Omega_p$, $\Omega_n$, and $K$ will be constants of the motion; and as with the odd-$A$ case, since then $L_3 = 0$, Eq. (1) yields

$$K = \Omega_p + \Omega_n. \tag{2a}$$

The Hamiltonian for these systems is generally

$$\begin{aligned} H &= H_{\text{rot}} + H_p + H_n + V_{pn} \\ &= \frac{\hbar^2}{2} \sum_{k=1}^{3} \frac{L_k^2}{\mathscr{I}_k} + \frac{P_p^2}{2m_p} + V_p + \frac{P_n^2}{2m_n} + V_n + V_{pn} \end{aligned} \tag{3}$$

and we shall set $V_{pn}$ equal to zero in what follows. Applying the condition of axial symmetry and using Eq. (1), we can again group the operators in the following way: into $H_R$ we place all terms diagonal in the total angular momentum and its body-axis projection; into $H_c$ we place all of the Coriolis-like terms, not only ones of the form $\mathbf{I} \cdot \mathbf{j}$, but also those of the form $\mathbf{j}_p \cdot \mathbf{j}_n$ that step the particle states; and finally into a term called $H_{p,n}$ we place all of the remaining operators, all of them particle operators. Thus this Hamiltonian becomes

$$H = H_R + H_{p,n} + H_c$$

where

$$H_R = \frac{\hbar^2}{2\mathscr{I}_0} [I(I+1) - K^2] \tag{4}$$

$$\begin{aligned} H_{p,n} &= \frac{p_p^2}{2m_p} + V_p + (j_{p_1}^2 + j_{p_2}^2) \\ &\quad + \frac{p_n^2}{2m_n} + V_n + (j_{n_1}^2 + j_{n_2}^2) \end{aligned} \tag{5}$$

$$H_c = -\frac{\hbar^2}{\mathscr{I}_0} [(I_1 j_{p_1} + I_2 j_{p_2}) + (I_1 j_{n_1} + I_2 j_{n_2}) - (j_{p_1} j_{n_1} + j_{p_2} j_{n_2})]. \quad (6a)$$

By again using the stepping operator definition

$$V^{\pm} = V_1 \pm iV_2$$

the Coriolis term becomes

$$H_c = -\frac{\hbar^2}{2\mathscr{I}_0} [(I^+ j_p^- + I^- j_p^+) + (I^+ j_n^- + I^- j_n^+) - (j_p^+ j_n^- + j_p^- j_n^+)]. \quad (6b)$$

The first and third terms step $K$ and $\Omega$ down whereas the second and fourth terms step $K$ and $\Omega$ up. The last two terms are diagonal in $K$ but step one $\Omega$ up and the other down. It is these last two terms which make a diagonal contribution. The others possess only zero diagonal matrix elements. As will become apparent, the diagonal contribution arises when the two nucleons in $\Omega = \frac{1}{2}$ bands couple to $K = 0$.

We shall return to the matrix elements of $H_c$ in due course; however, first let us assume that the two particles are bound loosely enough to the core so that their intrinsic level spacing will be quite large compared with the rotational level spacing. Thus, except for the special case where diagonal contributions of $H_c$ arise, we can neglect its effect and the rotational and particle equations separate, yielding a product state function. The Schrödinger equation is

$$\begin{aligned} H \mid EIMK = \Omega_n + \Omega_p\rangle &= (H_R + H_{p,n}) \mid EIMK = \Omega_n + \Omega_p\rangle \\ &= E_{I,K} \mid EIMK = \Omega_n + \Omega_p\rangle. \end{aligned} \quad (7)$$

At this point it is necessary to consider how the odd particles couple to form the $K$ band. Because of the nuclear symmetry the particle states are degenerate with respect to the sign of $\Omega$ so that Eq. (2a) is more properly written

$$K = \mid \Omega_p \pm \Omega_n \mid. \quad (2b)$$

In this representation the solution of Eq. (7) is simply

$$E_{I\,K} = \frac{\hbar^2}{2\mathscr{I}_0} [I(I+1) - K^2] + E_{\Omega_p} + E_{\Omega_n}. \quad (8a)$$

Now since a band head has $I = K$, a pair of particle states will form two rotational bands of the form

$$\begin{aligned} I_1 &= | \Omega_p - \Omega_n |, \ | \Omega_p - \Omega_n | + 1, \ | \Omega_p - \Omega_n | + 2, \ldots \\ I_2 &= \Omega_p + \Omega_n, \ \Omega_p + \Omega_n + 1, \ \Omega_p + \Omega_n + 2, \ldots \end{aligned} \tag{9}$$

and with heads at

$$E_{I,K_1} = \frac{\hbar^2}{2\mathscr{I}_0} | \Omega_p - \Omega_n | + E_{\Omega_p} + E_{\Omega_n} \tag{10a}$$

$$E_{I,K_2} = \frac{\hbar^2}{2\mathscr{I}_0} (\Omega_p + \Omega_n) + E_{\Omega_p} + E_{\Omega_n}. \tag{10b}$$

Therefore the two associated rotational bands would be separated by

$$\Delta E = \frac{\hbar^2}{\mathscr{I}_0} \Omega_< \tag{10c}$$

where $\Omega_<$ is the smaller of $\Omega_p$ and $\Omega_n$. For the ground state band the relations (10) are not always satisfied and the question then arises which of the two possibilities of Eq. (2b) forms the ground state. A similar situation arises in the independent particle shell model of odd-odd-$A$ nuclei and it is resolved by a well-known set of empirical rules due to Nordheim [136]. Gallagher and Moszkowski [98] have extended these relations to deformed nuclei in the following manner. They assumed that the intrinsic spins of the odd neutron and proton, respectively, $\Sigma_n$ and $\Sigma_p$ (recall the relation $\Omega = \Lambda + \Sigma$) always couple parallel. From this assumption we obtain the coupling rules

$$\begin{aligned} I_0 &= | \Omega_p - \Omega_n |; \qquad \Omega_p = \Lambda_p \pm \tfrac{1}{2}, \ \Omega_n = \Lambda_n \mp \tfrac{1}{2} \\ I_0 &= \Omega_p + \Omega_n; \qquad \Omega_p = \Lambda_p \pm \tfrac{1}{2}, \ \Omega_n = \Lambda_n \pm \tfrac{1}{2}. \end{aligned} \tag{11}$$

It has been shown that these rules account in general for the ground state spins of deformed odd-odd nuclei [98].

In order to return to the energy eigenvalue problem we must consider next the matrix elements of $H_c$ and for this we need an explicit expression for the symmetrized state functions. Again we use the coupled particle representation

$$| \Omega \rangle = \sum_j C_{j\Omega} | j\Omega \rangle = \sum_j C_{j\Omega} \chi_{j\Omega}.$$

Thus the symmetrized and normalized state functions are

$$| EIMK = | \Omega_p \pm \Omega_n |\rangle = \sqrt{\frac{2I+1}{16\pi^2}} \sum_{j_n, j_p} C_{j_p\Omega_p} C_{j_n K \mp \Omega_p}$$

$$\times [\chi^{(1)}_{j_p\Omega_p} \chi^{(2)}_{j_p K \mp \Omega_p} D^{I*}_{MK} + (-1)^{I-j_p-j_n} \chi^{(1)}_{j_p-\Omega_p} \chi^{(2)}_{j_n - K \pm \Omega_p} D^{I*}_{M-K}] \quad (12)$$

using Eq. (4-6) *et seq.* The special case when $K = 0$ is of some interest, for then the state function becomes

$$| EIMK = 0\rangle = \sqrt{\frac{2I+1}{16\pi^2}} \sum_{j_n, j_p} C_{j_p\Omega_p} C_{j_n-\Omega_p}$$

$$\times [\chi^{(1)}_{j_p\Omega_p} \chi^{(2)}_{j_n-\Omega_p} + (-1)^{I-j_p-j_n} \chi^{(1)}_{j_p-\Omega_p} \chi^{(2)}_{j_n\Omega_p}] D^{I*}_{M0} \quad (13)$$

Therefore, the rotational energy level structure of a band with $K = 0$ will not possess the usual rotational spectrum. However, the even spin and odd spin levels will separate and will be displaced, one relative to the other. Thus, the energies may be written

$$E_{I,K=0} = E_{\text{rot}}(I) + E^{(\text{even})}_{\text{Pair}}, \qquad I \text{ even}$$
$$= E_{\text{rot}}(I) + E^{(\text{odd})}_{\text{Pair}}, \qquad I \text{ odd} \quad (14)$$

and in fact displaced $K = 0$ bands like these have been observed. In $Ho^{166}$ the ground state band has $K = 0$ and is built upon proton and neutron intrinsic states $\frac{7}{2}-$ [523] and $\frac{7}{2}+$ [633], respectively, with the ground state being $0-$ (see Fig. 5-1) while the $1-$ state is displaced above the $2-$ state of that band. The opposite is true in $Am^{242}$ wherein the ground state, the $1-$ level of the $K = 0$ band, is built on the proton and neutron states $\frac{5}{2}-$ [523] and $\frac{5}{2}+$ [622]. (Note that this latter case satisfies the Gallagher–Moszkowski coupling rules of Eq. (11) whereas the former example does not.)

It is now possible to calculate the matrix elements of $H_c$ for which the most general expression is

$$\langle EIMK' = | \Omega_n' \pm \Omega_p' | \, | H_c | \, EIMK = | \Omega_n \pm \Omega_p |\rangle$$

$$= -\frac{\hbar^2}{2\mathscr{I}_0} \sum_{j_n, j_p} C^*_{j_p\Omega_p'} C_{j_p\Omega_p} C^*_{j_n\Omega_n'} C_{j_n\Omega_n}$$

$$\times [\{[\delta_{K', K-1}\, \delta_{\Omega_p, \Omega_p - 1} + (-1)^{I-j_p-j_n} \delta_{K', -(K+1)}\, \delta_{\Omega_p' - (\Omega_p - 1)}]$$

$$
\begin{aligned}
&\times [(I+K)(I-K+1)(j_p+\Omega_p)(j_p-\Omega_p+1)]^{1/2} \\
&+ \delta_{K',\,K+1}\delta_{\Omega_p',\,\Omega_p+1} \\
&\times [(I-K)(I+K+1)(j_p-\Omega_p)(j_p+\Omega_p+1)]^{1/2}\} \\
&\times \delta_{\Omega_n',\,K'-\Omega_p'}\,\delta_{\Omega_n,\,K-\Omega_p} \\
&+ \{[\delta_{K',\,K-1}\delta_{\Omega_p',\,\Omega_p}\delta_{\Omega_n,\,K\pm\Omega_p-1} \\
&+ (-1)^{I-j_p-j_n}\delta_{K',\,-(K-1)}\delta_{\Omega_p',\,\Omega_p}\delta_{\Omega_n,\,-(K\pm\Omega_p-1)}] \\
&\times [(I+K)(I-K+1)(j_n-\Omega_n)(j_n+\Omega_n+1)]^{1/2} \\
&+ \delta_{K',\,K+1}\delta_{\Omega_p',\,\Omega_p}\delta_{\Omega_n,\,K\pm\Omega_p+1} \\
&\times [(I-K)(I+K+1)(j_n+\Omega_n)(j_n-\Omega_n+1)]\} \\
&\times \delta_{\Omega_n',\,K'\pm\Omega_p'} - \{[\delta_{\Omega_p',\,\Omega_p+1}\delta_{\Omega_n,\,K-\Omega_p-1} \\
&+ (-1)^{I-j_p-j_n}\delta_{K,0}\delta_{\Omega_p',\,-(\Omega_p+1)}\delta_{\Omega_n-(K-\Omega_p-1)}] \\
&\times [(j_p-\Omega_p)(j_p+\Omega_p+1)(j_n+\Omega_n)(j_n-\Omega_n+1)]^{1/2} \\
&+ [\delta_{\Omega_p',\,\Omega_p-1}\delta_{\Omega_n,\,K-\Omega_p+1} \\
&+ (-1)^{I-j_p-j_n}\delta_{K,0}\delta_{\Omega_p',\,-(\Omega_p-1)}\delta_{\Omega_n-(K-\Omega_p+1)}] \\
&\times [(j_p+\Omega_p)(j_p-\Omega_p+1)(j_n-\Omega_n)(j_n+\Omega_n+1)]^{1/2}\} \\
&\times \delta_{K\ K'}\delta_{\Omega_n',\,K'-\Omega_p'}]. \qquad (15)
\end{aligned}
$$

Here only the ultimate and antepenultimate terms possess diagonal contributions and then only if $|\Omega_p| = |\Omega_p'| = |\Omega_n| = |\Omega_n'| = \frac{1}{2}$. Explicitly these are

$$
\begin{aligned}
&\langle EIMK = 0 \mid H_c \mid EIMK = 0\rangle \\
&= (-1)^{I+1}\frac{\hbar^2}{2\mathcal{J}_0}a_p a_n \delta_{K,0}(\delta_{\Omega_p,\,1/2}\delta_{\Omega_n,\,-1/2} + \delta_{\Omega_p,\,-1/2}\delta_{\Omega_n,\,1/2}), \qquad (16)
\end{aligned}
$$

where $a_p$ and $a_n$ are, respectively, the proton and neutron decoupling parameters, defined in Eq. (4-8a)

$$a = \sum_j (-1)^{j+1/2}(j+\tfrac{1}{2}) \mid C_{j1/2}\mid^2.$$

Adding these diagonal contributions to the eigenvalue expression (8a) yields

$$
\begin{aligned}
E_{I,K} = \frac{\hbar^2}{2\mathcal{J}_0}&[I(I+1) - K^2 + (-1)^{I+1}a_p a_n\,\delta_{K,0} \\
&\times (\delta_{\Omega_p,\,1/2}\delta_{\Omega_n,\,-1/2} + \delta_{\Omega_p,\,-1/2}\delta_{\Omega_n,\,1/2})] + E_{\Omega_p} + E_{\Omega_n}. \qquad (8b)
\end{aligned}
$$

It should be noted in this expression that for the $K = 0$ bands, the last term in the brackets alternates in sign which will provide the beforementioned displacement of even spin and odd spin rotational levels for this particular type of $K = 0$ band. For this case there will be coupling by $H_c$ to the $K = 1$ bands that are formed by the opposite coupling of the $|\Omega| = \frac{1}{2}$ particle states.

A particularly good example of these various features of an odd-odd nucleus is seen in Fig. 5-1 which displays the observed energy levels of

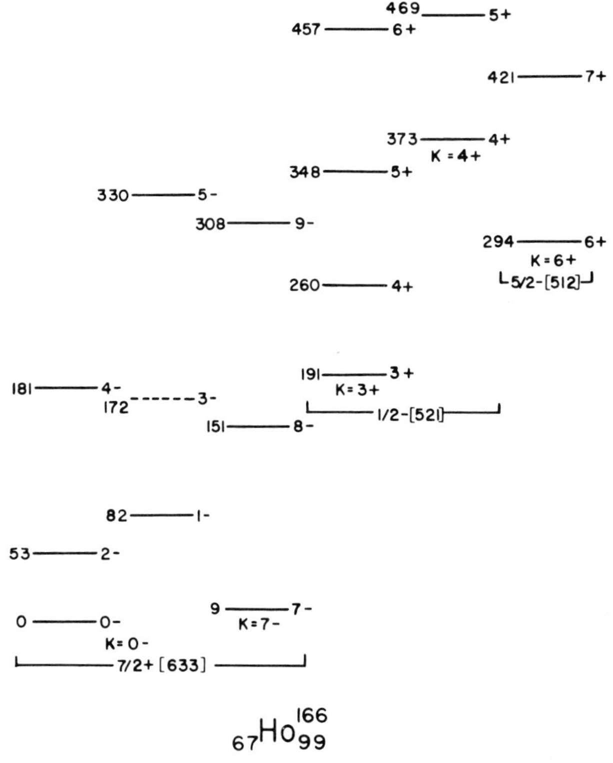


Fig. 5-1. The experimentally observed low-lying energy level structure of the odd-odd nucleus $Ho^{166}$. Each pair of bands formed from a particular intrinsic proton and neutron state are grouped together (except for a $K = 1+$ band which has not been observed). In the $K = 0$ ground state band the even and odd spin members have been displaced sideways in order to accentuate the even-odd spin effect of such bands. The intrinsic neutron level is noted below each pair of bands since they all arise from the coupling to the single proton state $\frac{7}{2} - [523]$. The position of the broken line shows a level that has been inferred theoretically. Energies, to the left of each level, are in keV.

the nucleus $Ho^{166}$ (the broken line depicts a level that is not observed but is inferred from theory). Here a single intrinsic proton state, the $\frac{7}{2}-$ [523], couples successively to three intrinsic neutron states—the $\frac{7}{2}+$ [633], $\frac{1}{2}-$ [521], and the $\frac{5}{2}-$ [512]—to form $K = 0-$, $7-$, $3+$, $4+$, and $6+$ bands (a $K = 1+$ band formed with the last intrinsic neutron state has not been observed).

Finally, it is necessary to mention the various possible core excitations which, perhaps more than in the odd-$A$ case, will serve to complicate the observed level structure. On each $K$ band there can be built a beta vibrational band which, at least superficially, should be identical with its ground state band. The moment of inertia should of course be somewhat larger. Indeed, to the expression for the energy levels (8b) we could add a vibrational term of the form

$$-b[I(I+1) - K^2 + (-1)^{I+1} a_p a_n \delta_{K,0} (\delta_{\Omega_p,1/2}\delta_{\Omega_n,-1/2} + \delta_{\Omega_p,-1/2}\delta_{\Omega_n,1/2})]^2$$

and from this predict the associated beta band head. However, this approach will probably prove no more productive than it has for the other classes of deformed nuclei. We should also expect to find, in general, two gamma bands with $K = K_g \pm 2$ (except where $K_g = 0$ or 1). No positive examples of either beta or gamma vibrational bands have been observed in odd-odd nuclei.

It can be seen that detailed investigations of deformed odd-odd nuclei should prove a productive area of study that should override the inherent experimental difficulties involved.

CHAPTER 6

# Electromagnetic Properties of the Models

## A. Introduction

By now it is fairly known that since the solutions of the Schrödinger equation possess the stationary property, a close correspondence between model eigenvalues and observed nuclear level structure is not a particularly good test of a model. The most stringent test lies with the model state functions. In this and ensuing chapters we shall examine the various predictions concerning nuclear transitions arising from the form of the model state functions.

The calculation of electromagnetic properties is the most straightforward, requires the fewest assumptions, and gives the most clear-cut predictions that are relatively easy to verify experimentally. Indeed the discovery of regions of nuclei with large static quadrupole moments led initially to postulating large nuclear deformations and it was not until some years later that the characteristic rotor spectrum was discovered in these nuclei. At the present time, strongly enhanced electric quadrupole transitions are considered to be the signature of nuclear deformation. Furthermore, since the form of the electromagnetic operators is known for all multipole orders, we shall first review them and use the operators associated with the lowest multipole orders in order to calculate the properties of even-even vibrational nuclei. These give considerable insight into a model that is often considered a limiting form of a deformed nucleus. After this we shall examine the electromagnetic properties of all of the deformed nuclear models, even as well as odd-$A$. In each case, we shall deal with these properties in increasing multipole order.

The most useful quantity in the discussion of electromagnetic transitions is the reduced matrix element $B(\lambda, \Delta L)$, which is defined from the transition probability per unit time for the emission of a photon of energy $\hbar\omega = \hbar ck$; and angular momentum $\lambda$ is

$$T(\lambda) = \frac{8\pi(\lambda + 1)}{\lambda[(2\lambda + 1)!!]^2} \frac{k^{2\lambda+1}}{\hbar} B(\lambda, \Delta L) \tag{1}$$

so that

$$B(\lambda, \Delta L) = \frac{1}{2L_i + 1} \sum_{M_i M_f} |\langle f \mid \mathscr{E}^L_{\lambda\mu} \mid i\rangle|^2. \tag{2}$$

The quantities $\mathscr{E}^L_{\lambda\mu}$ are the appropriate multipole operators defined in the laboratory and are spherical tensors of order $\lambda$, $\mu$.

The lowest multipole order is the electric monopole or $E(0)$ with selection rule $\Delta I = 0$, with no parity change. The energy of the transition is carried off either by a positron-electron pair (in light nuclei) or by an atomic electron emitted in internal conversion. If, respectively, the initial and final atomic electron state functions are represented by $\mid e_i\rangle$ and $\mid e_f\rangle$, the nuclear and electron radii by $r_n$ and $r_e$, and the nuclear density by $\varrho_n$, then the perturbation Hamiltonian for the coulomb interaction inducing these transitions is

$$H(L = 0) = -e \int_0^\infty \varrho_n \langle e_f \mid \frac{1}{r_n} - \frac{1}{r_e} \mid e_i\rangle \, d\tau_n. \tag{3}$$

It is customary in discussing these transitions to write the transition probability for electric monopole internal conversion as

$$T(0) = \Omega\varrho^2,$$

where $\Omega$, the atomic factor, is a function of the transition energy. It is available in graphic form [64]. The nuclear strength parameter is

$$\varrho = \sum_{p=1}^{Z} \int \varphi_f^* \left[\left(\frac{r_p}{R}\right)^2 - \sigma\left(\frac{r_p}{R}\right)^4 + \dots\right] \varphi_i \, d\tau_n, \tag{4a}$$

$\varphi$ being the nuclear state functions. Since estimates of the quantity $\sigma$ show that it is of the order 0.1 or less, most calculations only make use of the first term in the integrand. In evaluating this expression we replace the sum over the protons by an integral in the usual manner; however, in evaluating the volume integrals it is known [140] that we must include in the nuclear radius terms arising from the constant volume condition [Eq. (B5)].

The next multipole order of interest is the dipole† and we wish to show here that these collective models predict a zero transition probability. They are in fact observed to be small [36].

For electric dipole transitions the argument is quite simple. The electric dipole operator is proportional to the electric dipole moment $\mathbf{d} = q\mathbf{r}$. Since the collective models considered here assume the center of mass (and charge) to be at rest, the electric dipole transition matrix elements vanish identically to all orders in the expansion of the nuclear radius. Thus for even-even nuclei they should arise from single-particle transitions. The diagonal matrix elements representing the static moment vanish since parity is quite a good quantum number in nuclei [130]. (We specifically exclude here the high energy $E1$ transitions of the so-called giant resonance. These arise from charge oscillations and effectively exhaust the dipole sum rule. They will be discussed in Chapter 8.)

The case of the $M1$ transitions is somewhat different although for even-even nuclei they are forbidden. The operator for even-even nuclei and the cores of other nuclei is, in units of the nuclear magneton

$$\boldsymbol{\mu} = g_c \mathbf{L}$$

and $g_c$ is the collective gyromagnetic ratio. The components of the operator $\mathbf{L}$ are stepping operators in the angular momentum representation stepping the projection quantum numbers by one unit (Appendix A) but are diagonal in $\mathbf{L}^2$. Thus $\boldsymbol{\mu}$ can induce transitions only between states with the same value of $L$. For the vibrational model the matrix elements that are not diagonal vanish since they are associated with different surfon numbers. For deformed nuclei the rotor symmetry requirements are such that $\Delta K$ is even (for models of both parities, symmetric or asymmetric) and again off-diagonal matrix elements vanish.

† Magnetic-monopole, $M0$, transitions are also possible, the selection rules being $\Delta L = 0$ with parity change. This parity change prohibits the emission of a single conversion electron or a single positron-electron pair. Since then, the most probable $M0$ mode is the simultaneous emission of two particles the experimental detection of such transitions is extremely difficult. The most complete review of these transitions is to be found in Borisoglebskii [26].

For even-even nuclei the ground static magnetic dipole moment vanishes identically, whereas for excited states its measurement gives an unambiguous value of $g_c$. This collective gyromagnetic ratio has usually been taken simply as $Z/A$; however, experimentally they are less and often by significant amounts. Such deviations give an insight into the nature of the particle motion underlying the collective motions. In fact, these deviations have yielded to analysis involving the pairing interaction [135].

We next consider the higher electric transitions. Assuming a uniform charge distribution the operators are

$$T^L_{\lambda\mu} = \frac{3Ze}{4\pi R_0{}^3} \int r^{\lambda+2}\, Y^*_{\lambda\mu}(\theta,\ \varphi)\, dr\, d\Omega\,, \tag{5a}$$

the integral being over the nuclear volume. The most important of the electric transitions is the quadrupole although some enhanced electric octupole transitions are known.

The higher magnetic multipole operators are defined by

$$M^L_{l,m} = -\frac{i}{\lambda+1}\sqrt{\frac{4\pi}{2l+1}} \int (\mathbf{r}\times\boldsymbol{\nabla})\, r^l\, Y_{lm}(\theta,\ \varphi)\cdot \mathbf{j}(\mathbf{r})\, d\tau, \tag{5b}$$

where $\mathbf{j}(\mathbf{r})$ is the current density. For the vibrational model this gives for $\lambda$-order surfaces [159]

$$M^L_{l.m} = -\frac{3Zei}{8\pi c\lambda} R_0^{l+1}[l(l+2\lambda+1)(2\lambda-1)/(l+1)]^{1/2} C(l,\lambda-1,\lambda;000)$$
$$\times \sum_\nu (-1)^\nu \dot{\alpha}_{\lambda\nu} \alpha^*_{\lambda m-\nu} C(l\lambda\lambda;\ m,\ -\nu,\ m-\nu).$$

At present these are of little interest mainly because of the extreme difficulty in measuring them—only a handful of octupole moments have been determined.

## B. Vibrational Models

In calculating the various transition probabilities and static moments of the multipole operators discussed in the previous section we particularly wish to emphasize those transitions that are completely forbidden for the vibrational model. This is quite instructive since both this model and the rotational models of even-even nuclei have the same origins. Also the deformed regions are bounded by the so-called transitional regions for which the vibrational model serves as the simplest useful model.

The lowest multipole order is the electric monopole and if we write the nuclear strength parameter as

$$\varrho = \langle f \mid T_0 \mid i \rangle, \tag{4b}$$

then the electric monopole operator $T_0$ is obtained from Eq. (4a) by integrating over the collective coordinates

$$T_0 = \frac{3Ze}{4\pi}\left[\frac{4\pi}{5} + \sum_{\mu} \mid \alpha_{\lambda\mu} \mid^2 + \frac{5}{3}\,\mathscr{C}_{\lambda}\right]. \tag{6a}$$

The explicit form of $\mathscr{C}_{\lambda}$ is to be found in Eq. (B-4) without the $\lambda$ sums (that is, $\lambda = \lambda' = \lambda''$ not summed over) since we are dealing here with pure $\lambda$ surfaces. Clearly, the first term makes no contribution to $E0$ transitions whereas the last term does not if $\lambda$ is odd so that the operator is at least of second order.

Since it is simpler to evaluate matrix elements for this model in the number representation, we use the transformation (2-13) and obtain

$$T_0 = \frac{3Ze\hbar}{8\pi B_{\lambda}\omega_{\lambda}}\,[b^{+}_{\lambda\mu}b_{\lambda\mu} + b^{+}_{\lambda-\mu}b_{\lambda-\mu} + (-1)^{\mu}(b^{+}_{\lambda\mu}b^{+}_{\lambda-\mu} + b_{\lambda-\mu}b_{\lambda\mu}) + 1]. \tag{6b}$$

From the form of this equation it is clear that no transitions will occur between states that differ by an odd number of surfons even though this is allowed by the angular momentum selection rules (for example, for $\lambda = 2$ the $E0$ transition is prohibited from the $L = 2$, $N = 2$ state to the $L = 2$, $N = 1$ state even though $\Delta L = 0$). For the allowed $0 - 0$, $\Delta N = 2$ transition we find that

$$\varrho^2 = \frac{9(2\lambda + 1)Z^2e^2\hbar^2}{32\pi^2 B_{\lambda}^{\,2}\omega_{\lambda}^{\,2}}. \tag{7}$$

The next allowed transitions are the electric quadrupole. By setting $\lambda = 2$ in Eq. (5a) and integrating over the collective nuclear coordinates, the electric quadrupole operator is seen to be a first-order one; however, for $\lambda$-order surfaces the electric $2^\lambda$ are all of the same form

$$T^L_{\lambda\mu} = \frac{3ZeR_0^{\lambda}}{4\pi}\,\alpha_{\lambda\mu} \tag{8a}$$

to first order. In the number representation this becomes simply

$$T^L_{\lambda\mu} = \frac{3ZeR_0^{\lambda}}{4\pi}\sqrt{\frac{\hbar}{2B_\lambda\omega_\lambda}}\,(b_{\lambda\mu} + (-1)^\mu\, b^+_{\lambda\mu}) \tag{8b}$$

the matrix elements of which vanish unless $\Delta N = 1$.

Two transitions of greatest importance in the transitional regions are:

(i) $N = 1,\ \lambda \rightarrow N = 0,\ 0$

for which we find

$$B(E\lambda;\ 1\lambda \rightarrow 00) = \left(\frac{3ZeR_0^{\lambda}}{4\pi}\right)^2 \frac{\hbar}{2\sqrt{B_\lambda C_\lambda}} \tag{9}$$

for the transition from the first excited state while the other is the "stop-over" transition from the second to first excited state

(ii) $N = 2,\ \lambda \rightarrow N = 1,\ \lambda,$

which is possible only for $\lambda$ even. Here the reduced matrix element is readily seen to be

$$B(E\lambda;\ 2\lambda \rightarrow 1\lambda) = \left(\frac{3ZeR_0^{\lambda}}{4\pi}\right)^2 \frac{\hbar}{\sqrt{B_\lambda C_\lambda}}. \tag{10}$$

From the selection rules on the surfon number the "cross-over" transition is forbidden in lowest order. Here the model contradicts experiment, for these cross-over transitions for $\lambda$ equal to 2 are observed in all even-even nuclei in the transition regions. We might say that the model is supported since the observed transitions are weak, unlike most collective $E2$ transitions. Thus we might carry out the calculation using the $E\lambda$ operator to second order. However, a second model prediction flowing from the surfon selection rule is that the electric $2^\lambda$-moments are identically zero (a meaningful statement only for $\lambda$ even). For the case of the low-lying levels of even-even nuclei, $\lambda$ is equal to 2 and we should expect no state to have a static, electric quadrupole moment. Recently, it has become possible to measure such moments of excited states and the nucleus $Cd^{114}$—long thought to be an excellent example of a vibrational nucleus—has been found to have a quadrupole moment of its first excited state of the order of a barn. A clear-cut test like this of the model shows that it fails when applied to this nucleus.

A third transition, which is interesting, is that from the first excited $0+$ state to the first excited state for which

$$B(E\lambda;\ 20 \rightarrow 1\lambda) = \left(\frac{3ZeR_0^{\lambda}}{4\pi}\right)^2 \frac{\hbar}{\sqrt{B_\lambda C_\lambda}}$$

identical with (10). Comparing this expression with Eq. (7), we note that the ratio

$$\left[\frac{\varrho(\lambda)Ze^2R_0^{2\lambda}}{B(E\lambda\colon 20 \rightarrow 1\lambda)}\right]^2 = (2\lambda + 1)8\pi \tag{11}$$

is dimensionless and would be a useful quantity to compare with experiment. (This is considerably different from a similar ratio $X$ given in [140]. In that paper the operator is linear in the vibrational parameter, while here it is quadratic.)

## C. Models of Even–Even Nuclei

We again calculate the reduced transition matrix elements for transitions between various states of the rotational model. Even though it is possible to make the discussion as general as before, we shall restrict ourselves to quadrupole and octupole surfaces as was done in Chapter 3. Necessary extensions are straightforward.

First we consider electric monopole transitions and since the surface is defined in the body-fixed system by Eq. (2-2) we make use of transformation (2-19), subject to the condition (2-20). The monopole operator defined in Eq. (4b) is found to be

$$T_0 = \frac{3ze}{4\pi}\left(\frac{4\pi}{5} + \beta_\lambda^2 + \frac{5}{3}\beta_\lambda^3 T_\lambda\right), \tag{12}$$

where

$$T_\lambda = \frac{1}{7}\sqrt{\frac{5}{\pi}}\cos 3\gamma, \qquad \lambda = 2$$

$$T_\lambda = 0, \qquad \lambda \text{ odd.}$$

The first term in Eq. (12) makes no contribution, the second term induces transitions between beta (zeta) band and ground state band, and the third term induces transitions between the gamma and ground state band for the quadrupole case.† Since this last term is identically zero for the octupole case, no $E0$ transitions can occur between the analog to the gamma band and the ground state band. These selection rules are identical with those of the vibrational case. Since the inclusion of "gamma" vibrations [that is, not taking $\varepsilon_\lambda \equiv 0$ in Eq. (2-21)] leads to the vibrational system, we again include, besides rotations, only deformation vibrations.

For these monopole transitions we need only discuss the case without axial symmetry. The results for the symmetric case are obtained by setting the asymmetry parameter to zero. The total state function is a solution of Eq. (3-35) and can be immediately separated (Appendix C, Section C).

$$| n_{\beta_\lambda} LM\pi\rangle = \Phi^{(\lambda)}_{LNn}(\beta_\lambda) \mid LM\pi\rangle$$

and of course the angular momentum parts yield unity in the matrix element. Thus it is necessary to calculate

$$\varrho^{(\lambda)} = \langle \Phi^{(\lambda)}_{LN'n'}(\beta_\lambda) \mid T_0 \mid \Phi^{(\lambda)}_{LNn}(\beta_\lambda)\rangle, \tag{13}$$

where the vibrational eigenfunctions $\Phi^{(\lambda)}_{LNn}(\beta_\lambda)$ are solutions of Eq. (3-36b). On using the definitions of Eq. (3-37) *et seqq.*, the matrix element (13) may be written

$$\varrho^{(\lambda)} = \delta_{NN'} \frac{3ze}{4\pi} N_{\nu_f} N_{\nu_i} \int_0^\infty D_{\nu_f}\left[\sqrt{2}\, Z_{1f}\left(\frac{\beta_\lambda}{\beta_f} - 1\right)\right]$$
$$\times \left(\beta_\lambda^2 + \frac{5}{3}\beta_\lambda^3 T_\lambda\right) D_{\nu_i}\left[\sqrt{2}\, Z_{1i}\left(\frac{\beta_\lambda}{\beta_i} - 1\right)\right] d\beta_\lambda. \tag{14}$$

† In this section $z$ is the atomic number so that it will not be confused with $Z$ the vibrational parameter introduced in Chapter 3, Section C.

Now for the case of transitions between beta and ground bands for which $\Delta I = 0$, then $Z_{1f} = Z_{1i} = Z_1$, $\beta_i = \beta_f = \beta(I, N)$ but $N_{\nu_f} \neq N_{\nu_i}$ as $\nu_f \neq \nu_i$. Indeed these normalization constants may be written

$$N_\nu^2 \int_0^\infty D_\nu^2\left[\sqrt{2}\ Z_1\left(\frac{\beta}{\beta(I, N)} - 1\right)\right] d\beta$$

$$= N_\nu^2 \frac{\beta(I, N)}{Z_1} \int_{-Z_1}^\infty D_\nu^2\left(\sqrt{2}\ y\right) dy = N_\nu^2 \frac{\beta(I, N)}{Z_1} I_\nu = 1. \tag{15}$$

In general, the integrals $I_\nu$ must be obtained by numerical methods [10].

It is customary to carry out calculations in these models only for the lowest nonvanishing order so that we may dispense with the $\beta_\lambda^3$ term in the integral of Eq. (14) which upon change of variable becomes

$$I^{(0)}_{\nu_f\nu_i} = \int_0^\infty D_{\nu_f}[\sqrt{2}\ (X - Z_1)]\ D_{\nu_i}[\sqrt{2}\ (X - Z_1)]X^2\, dX, \tag{16}$$

the 0 meaning monopole transitions since similar integrals arise for $E2$ transitions. With the definitions (15) and (16) and the fact that $Z = \beta(I,\ N)/\beta_0\mu$, the monopole nuclear strength parameter may be written

$$\varrho^2 = \left(\frac{3ze}{4\pi}\right)^2 \left(\frac{\mu Z\beta_{\lambda 0}}{Z_1}\right)^4 I^2_{\nu_f\nu_i}(0)/I_{\nu_i}I_{\nu_f}. \tag{17}$$

Since this expression cannot be given in closed analytic form it has been determined numerically both for quadrupole and octupole systems for several values of the angular momentum of the initial state [75]. In general the quantity $\varrho^2/Z^2\beta^4_{\lambda 0}$ is a very strong function of the stiffness parameter $\mu$ that increases monotonically with $\mu$, but is only weakly dependent upon the asymmetry parameters. For small $\mu$ this quantity is almost independent of the angular momentum wheareas for large values of $\mu$ ($\gtrsim 0.5$) it is more strongly dependent upon $L$ and shows the behavior

$$[\varrho^2/Z^2\beta^4_{20}]_{L_1} > [\varrho^2/Z^2\beta^4_{20}]_{L_2}$$

for $L_1 > L_2$.

The quantity $\varrho^2$ is certainly not the most useful one to compare with experiment since it contains too many free parameters that is, $\mu$, $\beta_{\lambda 0}$, etc.). In the literature we often find a comparison between the relative

$E0$ and $E2$ transition rates from a given level. This quantity is defined by [141]

$$\mu_K(L_i \to L_f) = \frac{T(E0: L \to L)}{T(E2: L_i = L \to L_f)}. \tag{18}$$

An even more useful quantity is the ratio defined for zero-zero transitions by [140]

$$X(O_\beta \to O_{\text{gnd}}) = \frac{\varrho^2 e^2 R_0{}^4}{B(E2: O_\beta \to \mathscr{T}_{\text{gnd}})} \tag{19}$$

and generalized to other monopole transitions [75]

$$X(I_\beta \to I_{\text{gnd}}) = \frac{\varrho^2 e^2 R_0{}^4}{B(E2: I_\beta \to I_{\text{gnd}})}. \tag{20}$$

The relation between the ratios $\mu_K$ and $X$ is simply

$$X = (2.53\ \mu_K A^{4/3} E_\gamma{}^5/\Omega) \times 10^9,$$

where $E_\gamma$ is the gamma transition energy in MeV.

By anticipating the discussion on $E2$ transitions we may write the $X$ ratios in rather simple form. For zero-zero transitions (19) becomes

$$\begin{aligned} X(0_\beta \to 0_{\text{gnd}}) = {} & (\mu\beta_{20})^2 (Z_i/Z_{1i})^5 (Z_{1f}/Z_f)^3 \\ & \times I_{\nu_f}(2) I_{\nu_f\nu_i}(0)^2 / I_{\nu_f}(0) I_{\nu_i\nu_f}(2)^2 b(E2 : 01+ \to 21+). \end{aligned} \tag{21}$$

Here the normalization integrals for the final states are different for each process, the (0) being associated with the monopole final state whereas the (2) is associated with the $E2$ final state. For other transitions this problem does not arise so that

$$\begin{aligned} X(I_\beta + \to I_{\text{gnd}} +) = {} & (\mu\beta_{20} Z_i/Z_{1i})^2 \\ & \times I_{\nu_f\nu_i}(0)^2 / I_{\nu_f\nu_i}(2)^2\, b(E2 : I_\beta N_\beta + \to I_{\text{gnd}} N_{\text{gnd}} +). \end{aligned} \tag{22}$$

The $E2$ overlap integrals $I_{\nu_i\nu_f}(2)$ in both (21) and (22), similar to the monopole overlap integral, were defined in Eq. (16), and are discussed later [Eq. (35)]. For the case of monopole transitions between negative parity states, the $E0$ and $E2$ overlap integrals are identical; thus the $X$ ratio is especially simple

$$X(I_\zeta - \rightarrow I_{\text{gnd}} -) = 3\pi(\mu\zeta_0 Z/Z_1)^2/4b(E2 : I_\zeta N_\zeta - \rightarrow I_{\text{gnd}} N_{\text{gnd}} -), \quad (23)$$

where $\zeta_0 = \beta_{30}$. In Eqs. (21)–(23) we have used the quantity $b(E2: IN \rightarrow I'N')$ as defined with respect to the adiabatic (that is, pure rotational) reduced transition probability $B_a(E2: I_i \rightarrow I_f)$ [Eq. (29)] by

$$b(E2: IN \rightarrow I'N') = (4\pi/3zeR_0{}^2)^2 B_a(E2: IN \rightarrow I'N'). \quad (24)$$

For the case of vibrations about axial symmetry and in the limit of $\mu$ small, the expression (21) becomes

$$X(0_\beta \rightarrow 0_{\text{gnd}}) = 4\beta_{20}^2 ,$$

which has been derived in another way [140].

These results have been compared with experiment [2, 75, 140, 141]; it is rather curious that good agreement is obtained in the actinides (except for $Pu^{240}$ which is explained by the fact that the beta band head lies very near to the neutron energy gap [2]) but that very poor agreement is obtained in the rare earth region. It has been suggested that the cause of the difficulties here is the mixing of other collective modes (perhaps vibrational) with the beta vibrations in such a manner that the $E0$, but not the $E2$, transitions are muted [74]. This question is still not resolved.

Since no adequate nonperturbation treatment of gamma vibrations has been made for the asymmetric model we shall not discuss the calculations of monopole transitions from this band [81]. The experimental situation is not altogether clear since these transitions, where observed, are smaller than similar transitions from the beta band. In general, fewer nuclei have measured monopole transitions from the gamma band than from the beta band.

As we have noted earlier, low energy, collective $E1$ transitions are prohibited by the model so that any $E1$ transitions that do occur will be single particle in nature. Indeed, since these transitions must break a pair, $E1$ transitions in the deformed regions should be strongly inhibited. Furthermore, from the tensor form of the electromagnetic operators $\mathscr{E}^L_{\lambda\mu}$ of Eq. (2), all gamma-ray transitions are subject to a $K$ selection rule

$$\Delta K \leq \lambda ,$$

which would be absolute if $K$ were a good quantum number. The degree of $K$-forbiddenness $\nu$ is defined as

$$\nu = \Delta K - \lambda$$

and gives a measure of the $K$-band mixing. In general, the larger $\nu$, the more inhibited is the transition. An example of a strongly inhibited $E1$ transition is that from the isomeric 9 − state at 1142.9 keV in $Hf^{180}$ to the 8 + state at 1085.3 keV. This latter state belongs to the $K = 0$ ground state rotational band whereas the former state, since it is the lowest negative parity state in $Hf^{180}$, belongs to a $K = 9$ band. Although the transition is allowed by the usual gamma-ray selection rules it is strongly $K$-forbidden with $\nu = 8$ and is much slowed.

We now turn to electric quadrupole transitions. The transition operator is defined in the body-coordinate system as

$$T^B_{2\nu} = \frac{3ze}{4\pi R_0{}^3} \int r^4 Y^*_{2\nu}(\theta, \varphi)\, dr\, d\Omega \tag{5c}$$

for a uniform charge distribution. This is related to the operator referred to the laboratory system by

$$T^L_{2\mu} = \sum_\nu D^{2*}_{\mu\nu}(\theta_i)\, T^B_{2\nu}\, ;$$

thus the transition matrix elements have the form

$$\begin{aligned} \langle f \mid T^L_{2\mu} \mid i\rangle = \frac{3ze}{4\pi R_0{}^3} \sum_\nu (-1)^{\mu-\nu} \langle L_f M_f \mid D^2_{-\mu-\nu}(\theta_j) \mid L_i M_i\rangle \\ \times \int r^4\, Y^*_{2\nu}(\theta, \varphi)\, dr\, d\Omega. \end{aligned} \tag{25}$$

Since all of the rotational models discussed in Chapter 3 have only even $K$ bands, $\nu$ can only be even and Eq. (25) becomes

$$\begin{aligned} \langle f \mid T^L_{2\mu} \mid i\rangle = \frac{3ze}{4\pi R_0{}^3} (-1)^\mu \{ \langle L_f M_f \mid D^2_{-\mu 0}(\theta_j) \mid L_i M_i\rangle \int r^4 Y^*_{20}(\theta, \varphi) dr\, d\Omega \\ + \frac{1}{2} \langle L_f M_f \mid D^2_{-\mu 2}(\theta_j) + D^2_{-\mu-2}(\theta_j) \mid L_i M_i\rangle \\ \times \int r^4 [Y^*_{22}(\theta, \varphi) + Y^*_{2-2}(\theta, \varphi)]\, dr\, d\Omega \\ - \frac{1}{2} \langle L_f M_f \mid D^2_{-\mu 2}(\theta_j) - D^2_{-\mu-2}(\theta_j) \mid L_i M_i\rangle \\ \times \int r^4 [Y^*_{22}(\theta, \varphi) - Y^*_{2-2}(\theta, \varphi)]\, dr\, d\Omega \}. \end{aligned} \tag{26}$$

The last term is zero since the rotational part of the matrix element

$$\langle L_f M_f \mid D^2_{-\mu 2}(\theta_j) - D^2_{-\mu -2}(\theta_j) \mid L_i M_i \rangle$$

vanishes for both the $A$ and $B_1$ representations (for even-parity and odd-parity models).

By introducing the quadrupole moment tensor

$$Q_\mu = \frac{3ze}{4\pi R_0{}^3} \sqrt{\frac{16\pi}{5}} \int r^4 Y^*_{2\mu}(\theta, \varphi)\, dr\, d\Omega \tag{27}$$

and through it the quantities

$$Q_{\mu 0} = \sqrt{\frac{5}{64\pi}}\, (Q_\mu + Q_{-\mu}), \tag{28a}$$

Eq. (26) may be more simply written as

$$\begin{aligned}
\langle f \mid T^L_{2\mu} \mid i \rangle &= (-1)^\mu \{ \langle L_f M_f \mid D^2_{-\mu 0}(\theta_j) \mid L_i M_i \rangle Q_{00} \\
&\quad + \langle L_f M_f \mid D^2_{-\mu 2}(\theta_j) + D^2_{-\mu -2}(\theta_j) \mid L_i M_i \rangle Q_{20} \} \\
&= \pm (-1)^\mu \sqrt{\frac{2L_i + 1}{2L_f + 1}}\, C(L_i 2 L_f;\, -M_i,\, -\mu,\, -M_f) \\
&\quad \times \Bigg\{ \Bigg[ \pm \frac{A_0^{L_f} A_0^{L_i}}{4} [1 \pm (-1)^{L_f}][1 \pm (-1)^{L_i}] C(L_i 2 L_f;\, 000) \\
&\quad \pm (-1)^{L_f + L_i} \sum_{K_i > 0} A^{L_f}_{K_i} A^{L_i}_{K_i} C(L_i 2 L_f;\, K_i,\, 0,\, K_i) \Bigg] Q_{00} \\
&\quad + \Bigg[ \frac{(-1)^{L_i}[1 \pm (-1)^{L_f}]^2}{2\sqrt{2}} A_0^{L_f} A_2^{L_i} C(L_i 2 L_f;\, 2,\, -2,\, 0) \\
&\quad + \frac{(-1)^{L_f}[1 \pm (-1)^{L_i}]^2}{2\sqrt{2}} A_2^{L_f} A_0^{L_i} C(L_i 2 L_f;\, 0,\, 2,\, 2) \\
&\quad \pm (-1)^{L_i + L_f} \sum_{\substack{K_i > 0 \\ K_f > 0}} A^{L_i}_{K_i} A^{L_f}_{K_f} [C(L_i 2 L_f;\, K_i,\, 2,\, K_f) \\
&\quad + C(L_i 2 L_f;\, K_i,\, -2,\, K_f)] \Bigg] Q_{20} \Bigg\}
\end{aligned} \tag{29}$$

where the upper sign is for transitions between states of positive parity and the lower sign is for transitions between states of negative parity. After carrying out the integration in Eq. (29) and on using the definition (2-19), the quantities defined in Eq. (28a) become, for the two cases of interest

$$Q_{\mu 0} = \frac{3zeR_0^2}{8\pi} \beta_2(\varepsilon_{2\mu} + \varepsilon_{2-\mu}), \qquad (\pi +) \tag{28b}$$

$$Q_{\mu 0} = -\frac{3zeR_0^2}{4\sqrt{3\pi^3}} \beta_3^2 \sum_{\sigma,\varkappa} \varepsilon_{3\sigma}\varepsilon_{3\varkappa}[C(323;\ \varkappa,\ \mu,\ \sigma) + C(323;\ \varkappa,\ -\mu,\ \sigma)], \qquad (\pi -). \tag{28c}$$

If deformation vibrations are not included, then the $\beta_\lambda$ are fixed parameters and the adiabatic or pure-rotational reduced transition probability $B_a(E2)$ is obtained by substituting Eq. (29) into the expression (2), using either (28b) or (28c). The resulting expression is quite general and has been used to compare $E2$ branching ratios with experiment both for the asymmetric model and the symmetric model. For the latter case, each term alone accounts for a different transition. Transitions within the ground state rotational band ($K = 0$) are governed by just the first term of Eq. (29) whereas other intraband transitions are governed by the second term. Interband transitions, say between gamma and ground state bands, are governed by the third or fourth terms. Clearly, then, for this model the gamma-ray branching ratios are simply proportional to the squares of Clebsch–Gordan coefficients. Unfortunately, a simple model like this does not account too successfully for the ratios associated with a number of interband transitions. By introducing some band mixing it is possible to obtain closer agreement with experiment. For example, a simple mixing of the gamma and ground state bands requires the addition of a parameter to the theory that involves the ratio of the two components of the quadrupole moment tensor. We may define this as

$$z_\gamma = \varepsilon_\gamma Q_{00}/\sqrt{2}\, Q_{20}\,, \tag{30}$$

where $\varepsilon_\gamma$ is the amplitude of the $K = 2$ band in the ground state $K = 0$ band for $L = 2$ states. That is, the first excited state is taken to be

$$|\,21\rangle = \sqrt{1 - \varepsilon_\gamma^2}\,|\,2K = 0\rangle + \varepsilon_\gamma\,|\,2K = 2\rangle \tag{31}$$

and for the second $L = 2$ state

$$| 22\rangle = - \varepsilon_\gamma \,| 2K = 0\rangle + \sqrt{1 - \varepsilon_\gamma{}^2}\, | 2K = 2\rangle, \tag{31}$$

this mixing being presumed to arise from a perturbation term of the form

$$H' = k\left(\frac{L_1{}^2}{\mathcal{J}_1} - \frac{L_2{}^2}{\mathcal{J}_2}\right).$$

This simple band mixing then multiplies the reduced transition probabilities between pure $K$ bands by a factor $f(z_\gamma, L_i, L_f)$

$$B(E2\colon\ L_i \to L_f) = B_0(E2\colon L_i \to L_f)\, f(z_\gamma, L_i, L_f).$$

For example, the reduced branching ratio $B(E2\colon 22 \to 20)/B(E2\colon 22 \to 00)$ can be calculated at once from Eq. (29) assigning the coefficients $A_K{}^L$ from inspection of Eqs. (31). The result is

$$\frac{B(E2\colon 22 \to 20)}{B(E2\colon 22 \to 00)} = \frac{10}{7}\,\frac{(1 + 2z_\gamma)^2}{(1 - z_\gamma)^2}$$

to lowest order in $\varepsilon_\gamma$.

Although this band mixing is not identical with the $K$-band mixing that occurs naturally in the asymmetric top problem, it is equivalent in the sense that for small asymmetries $z_\gamma$ and $\gamma$ are functionally related. The solution of the top problem for $L = 2$ yields for the lowest state [78]

$$A_2{}^2 = \varepsilon_\gamma = (3 \sin\gamma \cos 3\gamma - \cos\gamma \sin 3\gamma)/N_1$$

with

$$N_1{}^2 = 2\sqrt{9 - 8\sin^2 3\gamma}\,[\sqrt{9 - 8\sin^2 3\gamma} + \sin\gamma\, \sin 3\gamma + 3\cos\gamma\, \cos 3\gamma].$$

In the limit of small $\gamma$ these yield at once

$$\varepsilon_\gamma = \frac{4}{3}\,\gamma^3 \tag{32}$$

whereas from Eq. (28b)

$$\frac{Q_{00}}{\sqrt{2}\, Q_{20}} = \cot\gamma \approx \gamma^{-1}.$$

Thus from Eq. (30)

$$z_\gamma = \frac{4}{3}\gamma^2.$$

In general there is no reason why the $\beta$ band cannot also mix with the ground state band so that in place of Eq. (31) we would have

$$|\, 21\rangle = |\, 2_{\text{gnd}}\, K = 0\rangle + \varepsilon_\beta \,|\, 2_\beta\, K = 0\rangle + \varepsilon_\gamma \,|\, 2_\gamma\, K = 2\rangle$$

to lowest order. The factors $f(z, L_i\, L_f)$ have been tabulated for all possible simple couplings between ground, beta, and gamma bands [124]. The relation between $\varepsilon_\gamma$ and the stiffness parameter may be determined in a similar, but more complicated, way to that used in obtaining Eq. (32).

The deformation ($\beta_\lambda$ —) vibrations are analytically simple to incorporate into the reduced transition probabilities even though quantitative results must be obtained by numerical means. We now assume that the quantities $\beta_\lambda$ are variables and not fixed parameters. The matrix elements (29) are then multiplied by a vibrational term

$$S_{\nu'\nu} = \langle \Phi^{(\lambda)}_{L'N'n'} \,|\, f(\beta_\lambda) \,|\, \Phi^{(\lambda)}_{LNn}\rangle, \tag{33}$$

where the operators are seen from Eq. (28b, c) to be

$$f(\beta_2) = \beta_2$$
$$f(\beta_3) = \beta_3{}^2 = \zeta^2.$$

The reduced transition probabilities are then just

$$B(E2\colon LNn \to L'N'n') = B_a(E2\colon LN \to L'N')\, S_{\nu_f\nu_i}\,.$$

On using the notation of Eq. (14) *et seqq.*, the vibrational matrix elements may be written

$$S^{1/2}_{\nu_f\nu_i} = N_{\nu_f} N_{\nu_i} \int_0^\infty D_{\nu_f}\!\left(\sqrt{2}\left[\frac{Z_{1f}}{\beta_f}\,\beta_\lambda - Z_{1f}\right]\right) \beta_\lambda{}^M,$$
$$\times\, D_{\nu_i}\!\left(\sqrt{2}\left[\frac{Z_{1i}}{\beta_i}\,\beta_\lambda - Z_{1i}\right]\right) d\beta_\lambda \tag{34}$$

where the exponent $M$ is

$$M = 1, \qquad \lambda = 2,\ \pi +$$
$$M = 2, \qquad \lambda = 3,\ \pi -.$$

If we define the ratio

$$R_z \equiv Z_{1i}Z_f/Z_{1f}Z_i\,,$$

then Eq. (34) becomes

$$S_{\nu_f\nu_i} = (Z_{1i}Z_{1f}/Z_iZ_fI_{\nu_f}I_{\nu_i})(Z_f/Z_{1f})^{2M+2}\,(\mu\beta_{\lambda 0})^{2M}I^2_{\nu_f\nu_i}(2)$$

with

$$I_{\nu_f\nu_i}(2) = \int_0^\infty D_{\nu_f}(\sqrt{2}\,[y - Z_{1f}])y^M\,D_{\nu_i}(\sqrt{2}\,[R_z y - Z_{1i})\,dy\,, \quad (35)$$

which must be evaluated numerically.

The vibrational contributions have been calculated [73] for various interband ($\Delta n \neq 0$) and intraband ($\Delta n = 0$) transitions. Comparison with experiment is quite favorable, in general being within experimental error both for transitions between positive parity states ($\lambda = 2$) and between negative parity states ($\lambda = 3$).

The static electric quadrupole moments are simply the diagonal matrix elements of the operator $T^L_{20}$ in the $M = L$ state. For a deformed nucleus the intrinsic quadrupole moment $Q_0$ (that is, the body-fixed value of the moment) is related to the value measured in the laboratory $\langle Q\rangle$ by evaluating Eq. (29) for $K = M = L$

$$\langle Q\rangle = \frac{L(2L-1)}{(L+1)(2L+3)}\,Q_0. \quad (36)$$

Hence the measured values of this moment for even-even nuclei are zero in the ground state. The expression for $Q_0$ is obtained by evaluating Eq. (27) which to second order yields

$$Q_0 = \frac{3ze}{\sqrt{5\pi}}\,R_0{}^2\beta(1 + 0.16\beta). \quad (37)$$

For a deformed nucleus this quantity is not zero and since it is difficult to measure excited state quadrupole moments $Q_0$ is usually obtained from coulomb excitation [1]. For a symmetric nucleus the excitation of the first excited state is from Eq. (29) just

$$B(E2\colon\ 00 \to 20) = \frac{5}{16\pi^2}\,Q_0{}^2, \quad (38)$$

thus with $|Q_0|$ evaluated experimentally the deformation parameter, but not its sign, can be extracted from the relation (37). A compendium of the experimental reduced transition probabilities (38) and the associated deformation parameters $\beta_2$ for even-even nuclei is available [40].

Once the deformation parameter has been determined the hydrodynamic moment of inertia can be calculated (Eq. 3-17 for the symmetric model). We may compare this with the experimental moment of inertia determined from the energy of the first excited state

$$\mathcal{J}_x = 3\hbar^2/E(21)$$

and with the rigid moment of inertia

$$\mathcal{J}_{\text{rig}} = \frac{2}{5} AMR_0^2(1 + 0.31\beta).$$

It is by now well known that $\mathcal{J}_x$ is in every case greater than the hydrodynamic value but less than rigid value. This is in fact due to the contributions arising from pairing forces [49].

We conclude this section with a short discussion of electric octupole transitions that occur between even- and odd-parity bands. The appropriate operator, referred to the body, is just

$$T^B_{3\nu} = \frac{3ze}{4\pi R_0^3} \int r^5 Y^*_{3\nu}(\theta, \varphi)\, dr\, d\Omega \tag{5d}$$

that must then be transformed to the laboratory system

$$T^L_{3\mu} = \sum_\nu D^{3*}_{\mu\nu}(\theta_i)\, T^B_{2\nu}.$$

Now these $E3$ transitions take place between states belonging to the $A$ and $B_1$ representation so that again $\nu$ is only even and the reduced transition probability contains the matrix elements

$$\begin{aligned}
\langle L_f M_f \pi + \mid T^L_{3\mu} \mid L_i M_i \pi - \rangle &= (-1)^\mu \frac{3ze}{4\pi R_0^3} \\
&\times \{\langle L_f M_f \pi + \mid D^3_{-\mu 0} \mid L_i M_i \pi - \rangle \int r^5 Y^*_{30}(\theta, \varphi)\, dr\, d\Omega \\
&+ \langle L_f M_f \pi + \mid D^3_{-\mu -2} + D^3_{-\mu 2} \mid L_i M_i \pi - \rangle \int r^5 [Y^*_{32}(\theta, \varphi) \\
&+ Y^*_{3-2}(\theta, \varphi)]\, dr\, d\Omega\}.
\end{aligned} \tag{39}$$

Again simplification results if the electric octupole moment tensor is introduced

$$\Omega_\mu = \frac{3ze}{4\pi R_0^3} \sqrt{\frac{16\pi}{7}} \int r^5 Y_{3\mu}^*(\theta, \varphi)\, dr\, d\Omega \tag{40}$$

with

$$\Omega_{\mu 0} = \sqrt{\frac{7}{64\pi}} (\Omega_\mu + \Omega_{-\mu});$$

thus (39) becomes

$$\begin{aligned}
&\langle L_f M_f \pi + \mid T_{3\mu}^L \mid L_i M_i \pi - \rangle \\
&= (-1)^\mu \sqrt{\frac{2L_i+1}{2L_f+1}}\, C(L_i 3 L_f;\, -M_i,\, -\mu,\, -M_f) \\
&\times \Bigg\{ \Bigg[ \frac{A_0^{L_f} A_0^{L_i}}{2} [1 + (-1)^{L_f}][1 - (-1)^{L_i}]\, C(L_i 3 L_f;\, 0,\, 0,\, 0) \\
&- (-1)^{L_i + L_f} \sum_{K_f > 0} A_{K_f}^{L_f} A_{K_f}^{L_i} C(L_i 3 L_f;\, K_f,\, 0,\, K_f) \Bigg] \Omega_{00} \\
&+ \Bigg[ \frac{A_0^{L_f} A_2^{L_i}}{2\sqrt{2}} [1 + (-1)^{L_f}]^2\, C(L_i 3 L_f;\, 2,\, -2,\, 0) \\
&+ \frac{A_2^{L_f} A_0^{L_i}}{2\sqrt{2}} [1 - (-1)^{L_i}]^2\, C(L_i 3 L_f;\, 0,\, 2,\, 2) \\
&- (-1)^{L_f + L_i} \sum_{\substack{K_f > 0 \\ K_i > 0}} A_{K_f}^{L_f} A_{K_i}^{L_i} [C(L_i 3 L_f;\, K_i,\, -2,\, K_f) \\
&+ C(L_i 3 L_f;\, K_i,\, 2,\, K_f)] \Bigg] \Omega_{20} \Bigg\}.
\end{aligned}$$

In order to evaluate the $\mathcal{Q}_\mu$ within the model it is necessary to treat the integration of Eq. (40) with some care. In fact, for surfaces of the form of Eq. (2-2) the limits on the radial integral should be from zero to the smaller of the quadrupole or octupole surfaces. However, integrating from the origin out to the mean surface yields for the octupole moment tensors (40)

$$\Omega_\mu = \frac{3zeR_0^3}{8\pi} a_{3\mu}, \tag{41}$$

which are first-order operators.

Calculations for $E3$ transitions, as outlined here, have been carried out [72] but only in the adiabatic limit. Whereas this has been sufficient to treat adequately what little experimental information is available it leaves open the question of how the effect of deformation vibrations is to be taken into account. The problem arises from the fact that the initial and final state functions contain different, and independent, vibrational parameters.

The reduced transition probabilities for higher multipole transitions can be calculated in a manner identical with the foregoing; however, few, if any, such transitions have been observed between the lower lying levels of deformed nuclei.

## D. Models of Odd-$A$ Nuclei

With the coupling of a particle to a deformed core the electromagnetic operators assume a slightly different form as the single-particle operators must be added to the collective operators of Section A. In the case of electric monopole transitions we can expect that the single-particle operator will give little or no contribution to the transition probability since its nonvanishing matrix elements connect states that are two shells away (that is, $\Delta N = 2$ since the transition requires no change of parity).

The experimental situation concerning electric monopole transitions is rather curious. At present there are only two odd-$A$ nuclei with known beta bands $Np^{237}$ and $U^{235}$ that have been identified principally by the hindrance factor (see Chapter 7) of a rather weak alpha decay to the state in question. In $U^{235}$ the $\mu_k$ ratio [Eq. (18)] has been measured and the value is rather close to what would be predicted from a calculation of a purely collective transition within the core [52]. This is consistent with such transitions in even-even actinide nuclei. In $Np^{237}$ the $E0$ transition is very strongly mixed with an $M1$ that is prohibited by the model if beta vibrations are principally core excitations.

The electric and magnetic multipole operators are now

$$T^L_{\lambda\mu} = \sum_j e_j r_j^{\lambda} Y_{\lambda\mu}(\theta_j, \varphi_j) + \frac{3}{4\pi} ZeR_0^{\lambda} \alpha^*_{\lambda\mu} \tag{42a}$$

$$M^L_{\lambda\mu} = \frac{e\hbar}{2mc} \sum_j \left[ g_s \mathbf{s} + \frac{2}{\lambda+1} g_l \mathbf{l} \right] \cdot \boldsymbol{\nabla}_j (r_j^{\lambda} Y_{\lambda\mu}(\theta_j, \varphi_j)) + \frac{e\hbar}{mc} \frac{1}{\lambda+1} g_c \int \mathbf{L}(\mathbf{r}) \cdot \boldsymbol{\nabla}(r^{\lambda} Y_{\lambda\mu}(\theta, \varphi))\, d\tau. \tag{42b}$$

The sums run over only the extra-core nucleons (of effective charge $e_j$) taking part in the transition. In Eq. (42a) a recoil contribution has been omitted in the sum (hence $e_j$) whereas only the lowest order collective operator is included. In fact, the core contributions of both types of operator (in each case the second term) are similar to those from even-even nuclei discussed in Section C before.

It is useful at this point to introduce a set of selection rules which, although rigorous in the limit of infinite distortion, are only approximately satisfied in actual nuclei. These selection rules involve the asymptotic quantum numbers defined in Chapter 4, Section B, and in general if they are not fulfilled they tend to slow or "hinder" a particular transition. They are quite easy to calculate for any multipole order and some of the more useful ones are given in Table 6-1.

The dipole transitions, both electric and magnetic, are very interesting since they tend to confirm the general details of the deformed single-particle state functions. From what has been said earlier these transitions have no core contribution so that whatever information they yield will concern the single-particle functions.

First consider the electric dipole transitions. A simple shell model would predict no low energy $E1$ transitions in the deformed regions (since these regions are far from shell edges); however, a number of low energy transitions like this are observed. Examination of Figs. D3–D5 in the deformed regions of odd particle numbers immediately shows that the model would predict a number of these transitions, since for large $N$ or $Z$ there are numerous level crossings for which $\Delta N = 1$ and $\Delta j \leq 1$. It is illuminating to compare the experimental transition probability with both the single-particle estimate (Weisskopf limit) and the deformed-core estimate (Nilsson limit) [36]. By defining the hindrance factor $H_w$ or $H_N$ as the ratio of experimental reduced transition probability to the similar theoretical quantity ($H = B(E1)\,\mathrm{exp}/B(E1)t$), then we find that the single-particle estimates for these transitions are many orders of magnitude larger than the experimental values. On the other hand, the Nilsson state functions yield values much closer to the observed ones. Indeed it has been shown [36] that if the initial and final deformation are permitted to be different they can be chosen such that $H_N = 1$. The reason for the diminution of the reduced transition probability in the deformed case flows from the very large cancellation rising from state function components of different $l$. It is quite easy to

TABLE 6–1

Asymptotic Selection Rules for Electromagnetic Transitions of Multipole Order $E1$, $M1$, $E2$, and $E3$

| Multipole | Operator | $\mu$ | $\Delta\pi$ | $\Delta K$ | $\Delta N$ | $\Delta n_z$ | $\Delta\Lambda$ |
|---|---|---|---|---|---|---|---|
| $E1$ | $rY_{1\mu}$ | $\pm 1$ | yes | 1 | $\pm 1$ | 0 | 1 |
| | | 0 | | 0 | $\pm 1$ | $\pm 1$ | 0 |
| $M1$ | $l_\mu$ | $\pm 1$ | no | 1 | $\begin{Bmatrix} 0 \\ 2 \end{Bmatrix}$ | 1 | 1 |
| | | | | 1 | $\begin{Bmatrix} 0 \\ 2 \end{Bmatrix}$ | $-1$ | 1 |
| | $s_\mu$ | $\pm 1$ | | 1 | 0 | 0 | 0 |
| | $l_\mu$, $s_\mu$ | 0 | | 0 | 0 | 0 | 0 |
| $E_2$ | $r^2 Y_{2\mu}$ | $\pm 2$ | no | 2 | $\begin{Bmatrix} 0 \\ \pm 2 \end{Bmatrix}$ | 0 | 2 |
| | | $\pm 1$ | | 1 | $\begin{Bmatrix} 0 \\ 2 \end{Bmatrix}$ | 1 | 1 |
| | | | | 1 | $\begin{Bmatrix} 0 \\ -2 \end{Bmatrix}$ | $-1$ | 1 |
| | | 0 | | 0 | $\pm 2$ | $\pm 2$ | 0 |
| | | | | 0 | 0 | 0 | $\pm 2$ |
| | | | | 0 | 0 | 0 | 0 |
| $E3$ | $r^3 Y_{3\mu}$ | $\pm 3$ | yes | 3 | $\begin{Bmatrix} \pm 3 \\ \pm 1 \end{Bmatrix}$ | 0 | 3 |
| | | $\pm 2$ | | 2 | $\pm 3$ | $\pm 1$ | 2 |
| | | | | 2 | $\pm 1$ | $\mp 1$ | 2 |
| | | $\pm 1$ | | 1 | $\pm 3$ | $\pm 2$ | 1 |
| | | | | 1 | $\pm 1$ | $\pm 2$ | 1 |
| | | 0 | | 0 | $\pm 3$ | $\begin{Bmatrix} \pm 3 \\ \pm 1 \end{Bmatrix}$ | 0 |
| | | | | 0 | $\pm 1$ | $\pm 1$ | 0 |

calculate these reduced transition probabilities and in fact the expression for an electric transition of multipolarity $\lambda$ is just

$$
\begin{aligned}
B(E\lambda\colon I_i\Omega_i = K_i \to I_f\Omega_f = K_f) &= e'^2 \Big\{ \sum_{j_i j_f} (-1)^{l_i+j_f} C^*_{j_f K_f} C_{j_i K_i} \\
&\times \left[\frac{(2l_i+1)(2j_f+1)(2\lambda+1)}{4\pi}\right]^{1/2} \langle N_f l_f \mid r^\lambda \mid N_i l_i \rangle\, C(l_i \lambda l_f;\, 000) \\
&\times W(l_i\,, j_i\,, l_f\,, j_f;\, \tfrac{1}{2},\, \lambda)[C(I_i \lambda I_f;\, K_i\,, K_f - K_i\,, K_f) \qquad (43) \\
&\times C(j_i \lambda j_f;\, K_i\,, K_f - K_i\,, K_f) \\
&+ (-1)^{I_i+j_i} C(I_i \lambda I_f;\, -K_i\,, K_f + K_i\,, K_f) \\
&\times C(j_i \lambda j_f;\, -K_i\,, K_f + K_i\,, K_f)] \Big\}^2 .
\end{aligned}
$$

The effect of large cancellations can be seen by making use of the appropriate single-particle components $C_{j\Omega}$ in Eq. (43) for the case in which $\lambda = 1$.

On the other hand, we may insist that the initial and final state deformations be the same. Then the $E1$ transitions divide into three classes: one with $\Delta K = K_f - K_i = 0$ for which the hindrance factor $H_N$ is close to unity (agreement between theory and experiment); another with $\Delta K = +1$ for which the hindrance factor is small, on the order of one tenth to one hundredth; and the third, with $\Delta K = -1$ for which the hindrance factor is yet smaller. Several attempts have been made to improve the theoretical predictions for the $\Delta K = \pm 1$ cases. One of the most straightforward of these attempts is to include the effects of the Coriolis coupling arising from $H_c$ (Chapter 4). This is often important since the intrinsic level spacings can be of the same order of magnitude as the rotational level spacings. Very good agreement is obtained for the nuclei studied [155] although no complete study of the effect has been made for all observed $E1$ transitions. Another important contribution in these two classes of transition arises from the mixing of one quasiparticle states with the Nilsson levels. This effect has been calculated for a number of nuclei [155] and again improves comparison with experiment. (For the $\Delta K = 0$ class of transitions this effect, arising from pairing, produces poorer agreement.)

Magnetic dipole effects in deformed nuclei possess some historical interest in that their diagonal matrix elements were among the first properties subjected to a detailed analysis of the model. The general $\lambda$th-order magnetic operator referred to the body system can be obtained from expression (42b) by the usual substitutions [159]

$$M_{\lambda\mu}^{B} = (-1)^{\mu+1}\mu_0[20\sqrt{3}\, g_c R_0^{\lambda-1} A(\lambda) \sum_{\nu} D_{\mu-\nu}^{\lambda}(\theta_i)(I_\nu - j_\nu)$$
$$- g(\lambda)\sqrt{4\pi\lambda}\, r^{\lambda-1} \sum_{\nu,\sigma} C(\lambda-1, 1, \lambda; \sigma-\nu, \nu, \sigma)$$
$$\times D_{\mu-\sigma}^{\lambda}(\theta_i) Y_{\lambda-1,\sigma-\nu} j_\nu], \tag{44a}$$

where

$$A(\lambda) = \sqrt{\frac{\lambda(2\lambda+1)}{\lambda+1}}\, C(2\lambda 2;\ 1, -1, 0) \sum_j C(\lambda 1 j;\ 0, 0, 0)$$
$$\times C(22j;\ 0, 0, 0)\, W(\lambda j 11;\ 1\lambda)\, W(22\lambda j;\ 12). \tag{44b}$$

$W$ stands for Racah coefficients whereas the total particle gyromagnetic ratio is calculated to be

$$g(\lambda) = \frac{1}{\lambda} \sum_{l,\sigma} a_{l,\lambda-\sigma}^2 \left[\sigma g_s + \frac{2g_l}{l+1}(\lambda - \sigma)\right]. \tag{44c}$$

Here the coefficients $a_{lm}$ are appropriate to the uncoupled representation $|Nl\Lambda\Sigma\rangle$ defined in Eq. (4-17).

For the dipole case a great simplification of (44) is possible. In order to calculate the dipole moments we may take the magnetic dipole operator as

$$\boldsymbol{\mu} = g_c \mathbf{L} + g_j \mathbf{j}, \tag{45}$$

where $g_c$ is usually taken as $Z/A$, but as noted before this value is often rather high. In any event, using the relation (4-1) Eq. (45) becomes

$$\boldsymbol{\mu} = g_c \mathbf{I} + (g_K - g_c)\mathbf{j}. \tag{46}$$

Since $K$ is equal to $\Omega$ in the model we define $g_j$ as being identical to $g_K$. The diagonal matrix elements of (46) are in units of the nuclear magneton

$$\langle \mu \rangle = g_c I + \frac{g_K - g_c}{I+1}[K^2 + (-1)^{I-1/2}(2I+I)a\, \delta_{K,1/2}], \tag{47}$$

where the "decoupling" parameter $a$ is defined in Eq. (4-8a). For the purposes of determining from experiment the magnetic dipole properties we must use two experiments in the same nucleus to determine the unknown quantities $g_K$ and $g_c$ unless we are willing to take for $g_c$ the value of a neighboring even-even nucleus. (If the band has $K = \frac{1}{2}$, then $a$ must be determined by still another experiment.) Clearly, it is best to use the reduced transition probability for $M1$ transitions.

In order to calculate the off-diagonal matrix elements of the operator $\boldsymbol{\mu}$, Eq. (45) must be replaced by

$$\boldsymbol{\mu} = g_c\mathbf{L} + g_l\mathbf{l} + g_2\mathbf{s} \; ; \tag{48}$$

then Eq. (46) is replaced either by

$$\boldsymbol{\mu} = g_c\mathbf{I} + (g_l - g_c)\,\mathbf{j} + (g_s - g_l)\mathbf{s} \tag{49a}$$

or by

$$\boldsymbol{\mu} = g_c\mathbf{I} + (g_s - g_c)\,\mathbf{j} + (g_l - g_s)\mathbf{l}. \tag{49b}$$

In either case, the first term connects no states with $\Delta I$ not equal to zero so it makes no contribution. The second term is most easily calculated in the coupled representation, whereas the third term is to be calculated in the uncoupled representation. In the usual notation

$$\begin{aligned}\langle I_f M_f K_f \mid j_\sigma{}^L \mid I_i M_i K_i\rangle &= \Gamma C(I_i\,1\,I_f;\; K_i\,,\; K_f - K_i\,,\; K_f) \\ &\times \sum_\nu (-1)^\nu \Big\{ (-1)^{I_i+I_f} \sum_j C^*_{jK_f} C_{jK_i} \sqrt{j(j+1)} \\ &\times C(j\,1\,j;\, K_i\,,\; K_f - K_i\,,\; K_f)\,(\delta_{\nu,K_f-K_i} + \delta_{-\nu,K_f-K_i}) \\ &- (-1)^{I_i+1/2} \frac{C(I_i\,1\,I_f\,;\,\frac{1}{2},\,-1,\,-\frac{1}{2})}{C(I_i\,1\,I_f;\,K_i\,,\,K_f - K_i\,,K_f)}\,\frac{a}{\sqrt{2}}\,(\delta_{\nu,1} + \delta_{\nu,-1})\delta_{K,1/2}\Big\}\end{aligned} \tag{50a}$$

$$\begin{aligned}\langle I_f M_f K_f \mid s_\sigma{}^L \mid I_i M_i K_i\rangle &= \sqrt{\frac{3}{4}}\;\Gamma C(I_i\,1\,I_f\,;\; K_i\,,\; K_f - K_i\,,\; K_f) \\ &\times \sum_\nu (-1)^\nu \Big\{ (-1)^{I_i+I_f} \sum_{l_i \Lambda_i} |a_{l_i\Lambda_i}|^2\, C(\tfrac{1}{2},1,\tfrac{1}{2};\,\Lambda_i - K_f, K_f - K_i, \Lambda_i - K_i) \\ &\times (\delta_{\nu,K_f-K_i} + \delta_{-\nu,K_f-K_i}) + (-1)^{I_i+l_i-1/2}\sqrt{\frac{2}{3}} \\ &\times \frac{C(I_i\,1\,I_f\,;\,\frac{1}{2}\,,\,-1,\,-\frac{1}{2})}{C(I_i\,1\,I_f;\,K_i\,,\,K_f - K_i\,,\,K_f)} \sum_{l_i} |\,a_{l_i 1}|^2\,(\delta_{\nu,1} + \delta_{\nu,-1})\delta_{K,1/2}\Big\},\end{aligned} \tag{50b}$$

where

$$\Gamma = \frac{(-1)^\sigma}{2} \sqrt{\frac{3(2I_i+1)}{4\pi(2I_f+1)}}\, C(I_i\, 1\, I_f;\, -M_i,\, -\sigma,\, -M_f). \quad (50c)$$

A great simplification arises for the case of intraband transitions; that is, $\Delta K = K_f - K_i = 0$. For this case, on using

$$g_K K = \langle \chi_K \mid g_l l_3 + g_s s_3 \mid \chi_K \rangle ,$$

$$B(MI\colon I_iK \to I_fK) = \frac{3}{4\pi} C^2(I_i\, 1\, I_f;\, K, 0, K)$$
$$\times \mid (-1)^{I_i+I_f}(g_K - g_c)K + (-1)^{I_i+1/2} d_{I_f}[(g_l - g_c)a$$
$$+ (g_s - g_l) \sum_{l_i} (-1)^{l_i - 1} \mid a_{l_i0} \mid^2]\, \delta_{K,1/2} \mid^2 \quad (51a)$$

in units of nuclear magnetons squared. The quantity $d_{I_f}$ in Eq. (51a) is defined by

$$d_{I_f} = \frac{1}{\sqrt{2}} \frac{C(I_i\, 1\, I_f;\, \tfrac{1}{2}, -1, -\tfrac{1}{2})}{C(I_i\, 1\, I_f;\, K, 0, K)}. \quad (51b)$$

Expressions for interband transitions are considerably more complicated but may easily be obtained from Eqs. (49a), (50a), and (50b). The quantity defined in Eq. (50c) yields only a factor $\frac{1}{4}$ in the reduced transition probability. Since the unknown magnetic dipole parameters may be taken to be $g_c$, $g_k$ (and if $K = \frac{1}{2}$ the decoupling parameter $a$) then these are most easily determined from two measurements: one for the ground state magnetic moment [Eq. (47)] and one from an interband $M1$ transition [Eqs. (51)]. If $K = \frac{1}{2}$, then the decoupling parameter is determined from a second interband $M1$ transition measurement. [In this case a model check is obtained by comparing the decoupling parameter so obtained with that obtained from the measured energy spectrum Eq. (4-7b).]

Unfortunately the selection rules that permit $M1$ transitions also allow $E2$ transitions to take place and since, in the latter, the core makes a major contribution the $\Delta I = 1$, *no*, transitions will contain a large amount of $E2$ admixture. The measurable quantity then is the ratio of $E2$ to $M1$ intensities usually denoted by $\delta^2$ and is defined by

$$\delta^2(I \to I-1) = \frac{T(E2\colon I, K \to I-1, K)}{T(M1\colon I, K \to I-1, K)}. \quad (52)$$

In order to evaluate this it is necessary to have an expression for the intraband, $\Delta I = 1$, reduced transition probability for the $E2$ transition. This is simply

$$B(E2: I_i \rightarrow I_f) = C^2(I_i 2 I_f; K, 0, K)\, Q_{00}^2$$
$$= \frac{5}{16\pi} C^2(I_i 2 I_f; K, 0, K)\, Q_0^2, \tag{53}$$

where $Q_{00}$ and $Q_0$ were defined before. They are most easily determined by yet another experiment involving coulomb excitation. (Another, although probably less accurate, method involves measuring the cross-over to stop-over transitions from the second rotational level in the band; that is, $I = I_{\text{gnd}} + 2$.) Thus we have from Eqs. (51) and (53)

$$\delta^2(I \rightarrow I-1) = \frac{3}{20}\left(\frac{m\omega Q_0}{e\hbar}\right)^2 \frac{K^2}{(I^2-1)}$$
$$\times \left| (g_K - g_c) K + (-1)^{I_i+1/2} d_{I_f}[(g_l - g_c)a + (g_s - g_l) \sum_{l_i} (-1)^{l_i-1} |a_{l_i0}|^2 \delta_{K,1/2} \right|^2. \tag{54}$$

This last expression is ambiguous with regard to sign and the latter has been defined as [23] the sign of

$$\frac{g_k - g_c}{Q_0}. \tag{55}$$

Intraband $E2$ transitions play an important role in these nuclei and numerous relevant intensity relations can be derived from Eq. (53). For example, for intensity ratios involving cross-over to stop-over decays mentioned before, we find

$$\frac{B(E2: I \rightarrow I-2)}{B(E2: I \rightarrow I-1)} = \frac{(I+1)(I^2 - 2K - K^2)}{2(2I-1)K^2}, \tag{56}$$

whereas coulomb excitation from the ground state should follow the ratio

$$\frac{B(E2: I_g = K_g \rightarrow I_g + 2, K_g)}{B(E2: I_g = K_g \rightarrow I_g + 1, K_g)} = \frac{2(I_g+1)}{I_g(2I_g+3)}. \tag{57}$$

Transitions between bands can take place; however, we must differentiate between two types of these interband transitions. In one, the transitions take place between levels of different intrinsic states. States such as these belong to different asymptotic quantum numbers of the particle but the core quantum numbers remain the same. In the other, although the intrinsic particle states remain the same, the core states change and thus we are dealing with transitions between beta and ground ($\Delta K_c = 0$) state bands or between gamma and ground ($\Delta K_c = 2$) state bands. For this latter case, we should expect the results of the previous section to apply since the transition operator affects only the core states. Thus the intrinsic states integrate to unity in the matrix elements. The former type of transition between different intrinsic states will, because of the state function symmetrization, contain two terms

$$B(\lambda: I_i K_i \to I_f K_f) = | C(I_i \lambda I_f;\, K_i\,, K_f - K_i\,, K_f) \langle K_f \mid O^B_{\lambda, K_f - K_i} \mid K_i \rangle$$
$$+ (-1)^{I_i + l_i + 1/2} C(I_i \lambda I_f;\, -K_i\,, K_i + K_f\,, K_f) \langle K_f \mid O^B_{\lambda, K_i + K_f} \mid -K_i \rangle |^2, \tag{58}$$

where $O^B_{\lambda\mu}$ is either of the operators of Eqs. (42) in the body axes. The second term here is less important since it makes a contribution only when $\lambda \leq K_i + K_f$, which is usually not the case.

For those cases in which only the first term of Eq. (58) contributes, then the intensity ratios for a series of transitions—all of which have the same initial intrinsic states and the same final intrinsic states—will be simply the ratio of the squares of the appropriate Clebsch–Gordan coefficients. This can be a very powerful method to use to assign the quantum numbers to various states in bands built upon different intrinsic levels.

Magnetic octupole moments $\langle \Omega \rangle$ are the next higher static multipole moments of which a few have been measured. As might be expected, the theory of such moments within the model, although straightforward, is somewhat involved. A self-consistent calculation using Nilsson uncoupled state functions has been carried out [159] and the magnetic octupole moment is

$$\langle \Omega \rangle = -\mu_0 \frac{I(2I-1)(2I-2)}{(2I+4)(2I+3)(I+1)} \left[ \frac{6}{7} g_c R_0^{\,2} + \frac{g(3)\hbar}{m\omega_0} G(I, N, \beta) \right]. \tag{59}$$

In Eq. (59) $R_0$ is the usual nuclear radius and $\hbar/m\omega_0$ is the square of the oscillator length and $\mu_0$ is the nuclear magneton. The quantity $g(3)$

is defined more generally in Eq. (44c) whereas the function $G(I, N, \beta)$ is [159]

$$
\begin{aligned}
G(I, N, \beta) = \sum_{L,\sigma} \Big\{ & a^2_{L,I-\sigma} \frac{3(2N+3)}{2(2L-1)(2L+3)} \\
& \times [(3I-2\sigma)L(L+1) - (I-\sigma)^2(5I-2\sigma) - (I-\sigma)] \\
& + a_{L+2,I-\sigma}\, a_{L,I-\sigma} \frac{3(5I-2\sigma)}{(2L+1)(2L+3)(2L+5)} \\
& \times [(2L+1)(2L+5)[(L+1)^2-(I-\sigma)^2][(L+2)^2-(I-\sigma)^2] \\
& \times (N-L)(N+L+3)]^{1/2} \Big\} \\
& + \sum_L \Big\{ a_{L,I-1/2}\, a_{L,I+1/2} \frac{3I(2N+3)}{(2L-1)(2L+3)} [(2L+1)^2 - 4I^2]^{1/2} \\
& + a_{L,I-1/2}\, a_{L+2,I+1/2} \frac{3}{4(2L+1)(2L+3)(2L+5)} \\
& \times [(2L+1)(2L+5)(N-L)(N+L+3)[(2L+3)^2-4I^2] \\
& \times (2L+2I+1)(2L+2I+5)]^{1/2} \\
& - a_{L,I+1/2}\, a_{L+2,I-1/2} \frac{3}{4(2L+1)(2L+3)(2L+5)} \\
& \times [(2L+1)(2L+5)(N-L)(N+L+3)[(2L+3)^2-4I^2] \\
& \times (2L-2I+1)(2L-2I+5)]^{1/2} \Big\} \\
& + (-1)^{N+1}\, 10 \sum_L \Big\{ a_{L,1}\, a_{L,2} \frac{6L(L+1)(2N+3)}{2(L-1)(2L+3)} \sqrt{(L-1)\,(L+2)} \\
& + a^2_{L,1} \frac{3(2N+3)L(L+1)}{4(2L-1)(2L+3)} - a_{L+2,1}\, a_{L,2} \frac{3(L+2)}{2(2L+1)(2L+3)(2L+5)} \\
& \times [(2L+1)(2L+5)(L-1)L(L+1)(L+3)(N-L)(N+L+3)]^{1/2} \\
& - a_{L+2,2}\, a_{L,1} \frac{3(L+1)}{2(2L+1)(2L+3)(2L+5)} \\
& \times [(2L+1)(2L+5)L(L+2)(L+3)(L+4)(N-L)(N+L+3)]^{1/2} \\
& + a_{L+2,1}\, a_{L,1} \frac{3}{2(2L+1)(2L+3)(2L+5)} \\
& \times [(2L+1)(2L+5)L(L+1)(L+2)(L+3)(N-L)(N+L+3)]^{1/2} \Big\} \delta_{I,3/2}\,.
\end{aligned}
$$

We might hope that when adequate experimental information is available in the deformed regions a detailed study of these static moments might yield a better understanding of the current distributions in such nuclei.

## E. Models of Odd–Odd Nuclei

In the last chapter we discussed the model of odd-odd nuclei in which the final neutron and the final proton move independently in a deformed Nilsson potential. We saw that this model is quite useful in predicting the level structure and now we shall see how successful it is in explaining the electromagnetic properties. Since deformed odd-odd nuclei are not stable, they are quite difficult to study by ordinary means such as coulomb excitation. Indeed, one of the most fruitful methods has been through the stripping reactions from a neighboring odd-$A$ isotope. Even though useful information concerning the energy level structure is obtained this way, only a paucity of electromagnetic transition data is available and support of the model must lean heavily on magnetic-moment data and beta-decay systematics (Chapter 7, Section B).

The electromagnetic multipole operators are simple generalizations of those for odd-$A$ systems [Eqs. (42)]

$$T^L_{\lambda\mu} = e_p r_p{}^\lambda Y_{\lambda\mu}(\theta_p, \varphi_p) + e_n r_n{}^\lambda Y_{\lambda\mu}(\theta_n, \varphi_n) + \frac{3}{4\pi} ZeR_0{}^\lambda \alpha^*_{\lambda\mu} \tag{60a}$$

$$\begin{aligned} M^L_{\lambda\mu} = \frac{e\hbar}{2mc} \Big\{ & \Big[g_s \mathbf{s} + \frac{2}{\lambda+1} g_l \mathbf{l}\Big]_p \cdot \boldsymbol{\nabla}_p [r_p{}^\lambda Y_{\lambda\mu}(\theta_p, \varphi_p)] \\ & + \Big[g_s \mathbf{s} + \frac{2}{\lambda+1} g_l \mathbf{l}\Big]_n \cdot \boldsymbol{\nabla}_n [r_n{}^\lambda Y_{\lambda\mu}(\theta_n, \varphi_n)] \\ & + \frac{2}{\lambda+1} g_c \int \mathbf{L}(\mathbf{r}) \cdot \boldsymbol{\nabla}[r^\lambda Y_{\lambda\mu}(\theta, \varphi)]\, d\tau \Big\}. \end{aligned} \tag{60b}$$

In (60a) $e_p$ and $e_n$ are the effective charge of proton and neutron and may be considered fitting parameters. It is quite clear from the form of Eqs. (60) that two-particle transitions are absolutely forbidden. That is, we can expect to see no transitions in which both the neutron *and* the proton intrinsic states change. This type of forbiddenness has been called "nonoverlap"-forbiddenness and is in addition to the more usual $K$-forbiddenness [97]. It is illuminating to calculate the reduced transition probability in a general way where the transition operator in the body-fixed axis system is denoted by $0^B_{\lambda\mu}(x)$ and is either of the particle $(x)$ operators (60) suitably transformed. Then we obtain

$$
\begin{aligned}
B(\lambda: I_iK_i \to I_fK_f) = {} & | \, C(I_i\lambda I_f;\ K_i\,,\ K_f - K_i\,,\ K_f) \\
& \times [\langle N_{p_f}\Omega_{p_f} \mid N_{p_i}\Omega_{p_i}\rangle \langle K_f \mid 0^B_{\lambda,K_f-K_i}(n) \mid K_i\rangle \\
& + \langle N_{n_f}\Omega_{n_f} \mid N_{n_i}\Omega_{n_i}\rangle \langle K_f \mid 0^B_{\lambda,K_f-K_i}(p) \mid K_i\rangle] \\
& + (-1)^{I_i+1/2} C(I_i\lambda I_f;\ -K_i\,,\ K_i + K_f\,,\ K_f) \\
& \times [(-1)^{l_{n_i}} \langle N_{p_f}\Omega_{p_f} \mid N_{p_i}\Omega_{p_i}\rangle \langle K_f \mid 0_{\lambda,K_i+K_f}(n) \mid -K_i\rangle \\
& + (-1)^{l_{p_i}} \langle N_{n_f}\Omega_{n_f} \mid N_{n_i}\Omega_{n_i}\rangle \langle K_f \mid 0^B_{\lambda,K_i+K_f}(p) \mid -K_i\rangle] \, |^2, \qquad (61)
\end{aligned}
$$

which is a generalization of Eq. (58). Not only does Eq. (61) display the usual $K$ selection rules, but the selection rules ("nonoverlap-forbiddenness") on the untransformed particle are clearly displayed. Furthermore, the nonvanishing terms of (61) will have contributions that are identical to those calculated for similar cases for the odd-$A$ systems in the previous section and need not be restated here. We can say then that to the extent that the extra core proton and neutron are well represented by Nilsson state functions we should expect no two-particle transitions. Where these transitions do occur (and they do, but always very hindered) they will clearly be an excellent method of determining the importance of two-particle transition operators.

Single-particle transitions will also be hindered if the nontransforming particle state functions are indeed the same in initial and final state but the deformation is considerably different. In this case, the overlap will not be complete. This effect should be quite clear in transitions from vibrational bands of one particle to the ground state band of the same state. Unfortunately, no such transitions have been observed and similarly no calculations of the effect have been done but they should be straightforward.

In classifying the electromagnetic transitions we again have several possibilities to consider. We can have transitions within a $K$-band which, as noted in the foregoing, will be similar to the odd-$A$ case. We can also have core transitions from a beta or a gamma band to the ground state band.

Finally, we turn to the static moments. Quite a number of ground state magnetic moments have been measured in these systems and about half as many quadrupole moments. Both of these moments are easily calculated from the model. We may simply take two terms of the form of Eq. (47): one for the neutron and one for the proton which requires a knowledge of not only $g_c$ but also of $g_i$ for each of the particles. It is probably somewhat more consistent to start with the operator definition (48) and thus obtain the relation

$$\langle\mu\rangle = \frac{I}{I+1}\left[g_c + \langle l_{p_3}\rangle + g_s{}^p\langle s_{p_3}\rangle + g_s{}^n\langle s_{n_3}\rangle\right]. \tag{62}$$

We may use the unquenched, particle gyromagnetic ratios

$$\begin{aligned} g_l{}^p &= 1, & g_l{}^n &= 0 \\ g_s{}^p &= 5.585, & g_s{}^n &= -\,3.826 \end{aligned} \tag{63}$$

which leaves open the question of what value to use for $g_c$. The simplest course is to take for the core gyromagnetic ratio that of the liquid-drop model $Z/A \simeq 0.40$, which is one way to account for the collective effects on the nuclear current [109]. A more consistent way, however, is to use the value of $g_c$ from a neighboring odd-$A$ or even-even nucleus, if known, since this should more closely approximate the effects of any residual interactions such as pairing effects. A calculation such as this produces results quite comparable with measured values of the ground state magnetic moments, especially where the value of the quadrupole moment is available to determine the deformation parameter. This latter relation is adequately given by Eq. (37) and is necessary in order to calculate the average values of the operators in (62).

In general, the closeness of the predictions of the odd-odd-$A$ magnetic moments to observed values along with the success of the energy eigenvalue predictions allows one to place a good deal of confidence in the model. In particular, the picture of independent motion in a deformed Nilsson potential seems to account for most of the extra-core particle correlations.

# CHAPTER 7

# Alpha and Beta Decay

## A. Alpha Decay

If gamma transitions are the most useful single experimental tool in unravelling the details of nuclear structure, then certainly alpha transitions stand next in the list for heavy deformed nuclei. They share with gamma rays their monoenergetic property and being charged are easy to detect; in addition their energy is easy to determine. Modern techniques using alpha-gamma and alpha-electron coincidence methods, particularly with solid state detectors, make possible the investigation of alpha fine structure where some groups possess an intensity of less than $10^{-5}$ that of the strongest group (see for instance Lederer et al. [120]). With techniques such as these it is now possible to study levels in the daughter nucleus of the order of one million electronvolts above the ground state. A good example of the power of these alpha-coincidence techniques is to be found in the discovery of the beta band in $U^{235}$—the first odd $A$ nucleus in which a vibrational band such as this was positively identified [52]. The major drawback to their more general use is that only a relatively few nuclei are alpha emitters. Since the actinides form one of the largest of these groups, the study of alpha radioactivity is probably the best single source of nuclear structure information in this deformed region beyond radium.

Theoretically the alpha-decay problem can be said to consist of two essentially independent problems. These are related to the formation of an alpha particle at the nuclear surface from the particle orbitals available and then the problem of the barrier penetration of this charged four-nucleon system after formation. Thus if $\lambda$ be the alpha-decay constant, then we know that it is simply the product of the frequency $f$ with which the alpha particle strikes the nuclear surface and the barrier penetration probability $P(\varepsilon)$

$$\lambda = fP(\varepsilon), \tag{1}$$

where the penetrability is a strong function of the energy of the alpha particle. As written, Eq. (1) assumes the alpha-decay process to be rather simple since it is a decay from the initial or ground state of the parent nucleus to the final state (presumably the ground state) of the daughter nucleus. We shall want to generalize this expression shortly in order to discuss alpha transitions to excited states of the daughter nucleus in order to be able to extract nuclear-structure information from the data. Now the frequency $f$ is quite simply related to the reduced width $\gamma^2$, which in turn is directly proportional to the alpha formation factor. This is a somewhat complicated function of the nucleon orbitals, the nucleon potentials, and residual interactions. A great deal of progress has been made recently in calculating the function from various particle models, notably the shell model, with the addition of pairing forces through the nuclear analog of the BCS theory of superconductivity. Since these calculations are outside of the somewhat arbitrary limits we have set for this discussion, we can but refer the reader to a recent review article that is confined almost completely to the latest progress in this field [32].

The barrier penetrability, on the other hand, is dependent not only upon the alpha-decay energy but also on the alpha particle-nucleus potential function. This potential is, without the nucleus, simply the coulomb potential for a spherical, or in our case a deformed, uniform charge distribution. In the discussion that follows we shall confine our attention to the calculation of the penetrabilities.

The calculation of alpha-decay penetrabilities was probably the first successful application of quantum mechanics to nuclear physics. In 1928 Gamow [100] and Condon and Gurney [65] independently explained the relation between the decay constant and decay energy by calculating the penetration of an alpha particle through a spherically symmetric barrier. An empirical relation had been obtained from a systematic analysis of alpha half-lives and energies sometime before this by Geiger and Nuttall [101]. This relation is now usually expressed as

$$\log t_{1/2} = A(z)\, E_{\text{eff}}^{1/2} + B(z), \tag{2}$$

where $E_{\text{eff}}$ is the effective decay energy that has been corrected for electron screening whereas $A(Z)$ and $B(Z)$ are empirical quantities. A table of $E_{\text{eff}}$, $A(Z)$, and $B(Z)$ is to be found in Perlman and Rasmussen [37].

This happy state of affairs, with theory explaining an empirical relationship which in turn was in very good agreement with experiment, did not last long. By 1930 alpha fine structure had been discovered [143] and was soon surmised to be associated with decay to excited states of the daughter nucleus [99]. Not until quite sometime later did Hill and Wheeler suggest that nuclear deformations might effect both the alpha-decay rate and the angular distribution of the emitted particles [108]. Somewhat earlier it had been shown that there existed a quite clear relationship between the mass number of the parent nucleus and the alpha-decay energy [137]. This energy increased as the mass decreased until a shell edge was crossed where there was a very rapid decrease in the decay energy. Away from shell edges, that is, in the now recognized deformed regions, the ground state-ground state transitions of even-even nuclei were quite well explained by the Geiger–Nuttall rule. The decay to excited states in even-even nuclei departed from the expression (2) and often to a considerable extent—so much so that it was quite clear that the angular momentum barrier was not the only factor entering into the process. For odd-$A$ nuclei the situation was even more complicated with ground to ground-state transitions often far outside the predictions of Eq. (2) but then a ground to excited state transition would conform to this relation. We shall return to these systems later.

The most useful quantity with which to analyze alpha fine structure is the hindrance factor $F(Z)$, the factor by which the observed alpha half-life is different from that calculated from Eq. (2)

$$\log F(z) = \log t_{1/2} - A(z) E_{\text{eff}}^{1/2} - B(z). \tag{3}$$

Another useful quantity is the relative hindrance factor $f(Z)$, the hindrance factor that is normalized to the ground state hindrance factor

$$f_l(z) = F_l(z)/F_{\text{gnd}}(z) \tag{4}$$

the reciprocal $c_l$ is the reduced transition probability. We shall also use quantities $b_l$ in which the $l$ dependence of the centrifugal barrier has been removed. These latter are defined by

$$b_l^2 = c_l \exp\left[\frac{l(l+1)}{\eta}\left(\frac{2\eta}{kR_0} - 1\right)^{1/2}\right] \tag{5}$$

where

$$\eta = 2(z-2)e^2/\hbar v. \tag{6}$$

Here $v$ is the alpha-particle velocity *after* penetration of the barrier and $k$ is the alpha-particle wave number.

We now turn to the solution of the problem of the penetration of the alpha particle through the potential barrier of the daughter nucleus. In order to keep the problem as general as possible we shall assume the alpha particle carries with it $l$ units of angular momentum and that the barrier may not be spherically symmetric. In the center-of-mass system of the daughter and alpha particle the Hamiltonian is

$$H = H_\alpha + H_{\text{daughter}} + H_{\text{int}}, \tag{7}$$

with $H_\alpha$ simply the kinetic energy of the alpha particle; $H_{\text{int}}$ is the (deformed) electrostatic potential between the two systems and

$$H_{\text{daughter}} = H_{\text{rot}} + H_{\text{vib}} + H_{\text{part}}. \tag{8}$$

The three terms here are already familiar, having been discussed in Chapters 3 and 4. If we let $i$ stand for the initial (parent) quantum numbers and $f$ for the final (daughter) quantum numbers then the eigenfunctions associated with Eq. (7) may be written as

$$|I_iM_i\rangle = \exp(-iEt/\hbar) \sum_{I_f l} r^{-1} f_{lI_f}^{I_i(r)} \Phi_{M_i}^{I_i\tau}(l, I_f, \theta, \varphi, \theta_i, \beta_\lambda). \tag{9}$$

The coordinates $(r, \theta, \varphi)$ are the daughter-alpha separation and the orientation of the alpha relative to a laboratory fixed-axis system. As usual the $\theta_i$ represent the Euler angles relating the orientation of the deformed daughter relative to the laboratory, $\beta_\lambda$ refers to the vibrational coordinates, and the quantum number $\tau$ refers to the vibrational band. For states in the ground-state rotational band, $\tau = 1$; for states in the beta band $\tau = 2$, whereas $\tau = 3$ indicates states in the "gamma" band. Actually "...no gamma vibrations have been observed in alpha decay" [119] so this value will not be used. Even though beta bands have been observed in alpha decay of both even-even and odd-$A$ systems, we shall neglect $H_{\text{vib}}$ in what follows and indicate how the vibrations can be treated. In Eq. (9) the function $\Phi$ is defined generally as

$$\Phi_{M_i}^{I_i\tau}(l, I_f, \theta, \varphi, \theta_i, \beta_\lambda) = \sqrt{\frac{2I_f+1}{16\pi^2}}\, \Gamma_{\nu_n}^{(\tau)}(\beta_\lambda)$$
$$\times \sum_{M_f m} A_{K_f}^{I_f\tau} B_{\Omega_f}^{j_f} [\chi_{j_f\Omega_f} D_{M_f K_f}^{I_f *}(\theta_i) + (-1)^{I_f - J_f} \chi_{j_f - \Omega_f} D_{M_f - K_f}^{I_f *}(\theta_i)]$$
$$\times C(I_f l I_i;\, M_f, m, M_i) Y_{lm}(\theta, \varphi). \tag{10}$$

In neglecting $H_{\text{vib}}$ in Eq. (8) we must set $\Gamma_{\nu_n}^{(\tau)}(\beta_\lambda) = 1$ for all $\lambda$. However, if we wish to include the effects of various vibrations characterized here by $\nu_n$, we may simply use for $\Gamma_{\nu_n}^{(\tau)}(\beta_\lambda)$ the appropriate functions from Chapter 3. That is, if we make use of the adiabatic approximation then

$$\Gamma_{\nu_n}^{(2)}(\beta_2) = f(\beta) \tag{11a}$$

and

$$\Gamma_{\nu_n}^{(3)}(\gamma) = g(\gamma), \tag{11b}$$

which are defined in Eqs. (3-27) and (3-28). On the other hand, an exact calculation of the influence of the beta vibrations, allowing for asymmetric deformations, would use from Chapter 3 Section C

$$\Gamma_{\nu_n}^{(2)}(\beta_2) = D_{\nu_n}(a_1\beta_2 + a_2), \tag{11c}$$

where $D_{\nu_n}$ satisfies Eq. (3-41) and the $a_i$ are defined there. To encompass decay to negative parity states within the discussion only requires associating $\Gamma_{\nu_n}^{\tau}(\beta_3)$ with vibrational functions that were defined in Chapter 3, Section D.

Since we will permit nonaxial deformations we limit the surface expansion of Eq. (2-2) to terms with $\lambda = 2$ so that the electrostatic potential can be expanded as

$$V(\mathbf{r}') = \frac{2(Z-2)e^2}{r} + \sum_{\mu=-2}^{2} V_\mu(\beta, \gamma, r) Y_{2\mu}(\theta', \varphi'), \tag{12}$$

where clearly the expansion parameters are obtained from

$$V_\mu(\beta, \gamma, r) = \frac{8\pi}{5r^3} \int r''^2 \varrho(r'') Y_{2\mu}(\theta'', \varphi'')\, d\mathbf{r}''$$
$$= \frac{8\pi}{5r^3} Q_\mu, \tag{13}$$

the integral being over the nuclear volume. If we do not neglect vibrations, Eq. (13) gives a recipe (similar to that encountered in our discussion of gamma transitions) for finding the $\beta$ and $\gamma$ dependence of $H_{\text{vib}}$.

The barrier penetration problem has almost always been solved by using the WKB approximation and, for deformed calculations, by a variation suggested by Christy that has proved very powerful [63]. This requires the solution of the extremal problem

$$\int_{P}^{P'} K(\mathbf{r}')\,ds = \text{minimum}, \tag{14}$$

where $P$ and $P'$ are two points within the anisotropic potential barrier, and the arc length connecting them is measured from $P$ to $P'$. The integral may be expressed as the sum of a spherically symmetric part and a nonsymmetric part [138]

$$K(\mathbf{r}') = K_0(r) + \Delta K(r, \theta', \varphi')$$

with

$$K_0(r) = k\sqrt{\frac{2\eta}{kr} - 1}$$

$$\Delta K(r, \theta', \varphi') = \frac{2\pi\eta}{5(Z-2)e^2r^3}\left(\frac{2\eta}{kr} - 1\right)^{-1/2} \sum_{\mu=-2}^{2} Q_\mu Y_{2\mu}(\theta', \varphi'),$$

the quantity $\eta$ having been defined in Eq. (6).

The solution of the problem (14) involves, in the general case of an asymmetric barrier, the solving of a set of coupled differential equations that define the path to minimize the integral. If we assume the anisotropic part of the potential to be small compared with the isotropic part, then for an adequate approximation we use for the path, $PP'$, that which was used to solve the spherically symmetric problem. The actual numerical results are very critically dependent upon the alpha-particle boundary conditions at the nuclear surface. Two possibilities exist. Either we take the alpha-particle wave function constant there or we may expand it in spherical harmonics. (The constant condition has been used for both axially and nonaxially symmetric potentials. See Davidson [28] for a more complete discussion.) Proceeding in the most general fashion, we take the boundary condition as [138]

$$\bar{\psi}_0(\theta', \varphi') = \psi_0 \sum_{\mu} \varepsilon_{2\mu} Y_{2\mu}(\theta', \varphi') + \psi_1. \tag{15}$$

Then either we may use the empirical data to predict the exact details of the boundary condition or we may set the $\varepsilon_{2\mu}$ equal to zero.

The solution of Eq. (14) then yields

$$f_{lI_f}^{I_i}(r) = R_0\psi_0 \frac{G_l(\mathscr{E}_{I_f}^{\tau}, r)}{G_l(\mathscr{E}_{I_f}^{\tau}, R_0)} \sum_{\substack{K_f, \Omega \\ l'\Omega'}} (-1)^{I_i - I_f + \Omega} \times C(I_i l I_f;\ K_f + \Omega,\ -\Omega,\ K_f)\, A_{K_f}^{I_f\tau} h_{l\Omega;l'\Omega'}\, \varepsilon_{l'\Omega'}.$$

Here the functions $G_l$ are the well-known coulomb solutions that are irregular at the origin; the energy parameter $\mathscr{E}_{I_f}^{\tau}$ is the sum of the kinetic energy of the alpha particle and the recoil energy of the daughter nucleus. The $\varepsilon_{l\Omega}$ are defined in Eq. (15) while the matrix elements $h_{l\Omega;\,l'\Omega'}$ are defined by

$$h_{l\Omega;l'\Omega'} = \sum_{nm} \int Y_{l\Omega}^{*}(\theta', \varphi')\, C_{nm} Y_{nm}(\theta', \varphi') Y_{l'\Omega'}(\theta'\varphi')\, d\Omega'$$

with

$$\sum_{nm} C_{nm} Y_{nm}(\theta, \varphi) = \exp\left[\frac{2\alpha_\mu}{5} \sqrt{\frac{kR_0}{2\eta}} \left(1 - \frac{kR_0}{2\eta}\right) (4\eta - kR_0) Y_{2\mu}(\theta', \varphi')\right].$$

Since the wave function of Eq. (9) is now completely determined, the transition probability per unit time for the penetration through the barrier of an alpha particle of angular momentum $l$, leaving the daughter nucleus in the state $(I_f,\ \tau)$, is just

$$P_{lI_f\tau}^{I_i} = \lim_{r\to\infty} v_{I_f\tau}\, | f_{lI_f}^{I_i}(r) |^2,$$

$v$ being the velocity of the alpha particle. The most general expression for this transition probability is

$$P_{lI_f\tau}^{I_i} = v_{I_f\tau} \left(\frac{R_0\psi_0}{G_0(\mathscr{E}_{I_f}^{\tau}, R_0)}\right)^2 \exp\left[-\frac{l(l+1)}{\eta} \sqrt{\frac{2\eta}{kR_0} - 1}\right] \times \left| \sum_{\substack{K_f\Omega \\ l'\Omega'}} (-1)^{\Omega} A_{K_f}^{I_f\tau}\, C(I_i l I_f;\ K_f + \Omega,\ -\Omega,\ K_f)\, h_{l\Omega;l'\Omega'}\, \varepsilon_{l'\Omega'} \right|^2.$$

This relation then permits us to calculate the reduced transition probability $c_l$ for any alpha transition in deformed nuclei. It is more usual to calculate the $b_l$ that is defined in Eq. (5). Thus for even-even nuclei we have for transitions to the ground state rotational band of the daughter nucleus

$$b_{l1} = \left| \frac{\sum_{K_f} A_{K_f}(C_{lK_f} + \sum_{\Omega'} \varepsilon_{2-\Omega'} h_{lK_f;2-\Omega'})}{C_{00} + \sum_{\Omega'} \varepsilon_{2-\Omega'} h_{00;2-\Omega'}} \right|. \tag{16}$$

The empirical data permit a comparison of theory and experiment through $b_{81}$ only for $Pu^{238}$ and $Cm^{242}$. As is seen from Table 7-1a the agreement is adequate. In this table we also give the values of the deformation parameter $\beta_2$, the boundary condition parameters $\varepsilon_{2\mu}$ and the calculated values of $Q_\mu$. The values of both $\beta_2$ and $Q_\mu$ compare well with values determined by other methods. The quantities $\varepsilon_{2\mu}$ show that the alpha wave function is itself strongly distorted by the nuclear deformation. In general, the wave functions are greatly elongated at the equator and reduced at the poles. This supports an earlier suggestion [63] to explain the $l = 4$ intensity in the decay of $Cm^{242}$.

An alternative approach is to assume the nuclear surface to be symmetric and the alpha wave function constant thereon; but at the same time allow this surface more complexity so that Eq. (2-2) is replaced by [94]

$$R^B(\theta', \varphi') = R^B(\theta') = R_0[1 + \sum_{\lambda \geq 2}' \beta_\lambda Y_{\lambda 0}(\theta')], \tag{17}$$

the sum running only over even values of $\lambda$. This calculation has been carried out in exactly the same way in [94] with the terms through $\lambda = 8$ in Eq. (17) being kept. In Table 7-1b we list the deformation parameters $\beta_2$ and $\beta_4$ (for $Pu^{238}$ and $Cm^{242}$ the estimated values of $\beta_6$ and $\beta_8$ are about 0.02) as well as $b_{61}$ and $b_{81}$. The important point here, which has been confirmed by other studies [116] is that any second-order refinements to the theories of collective nuclear phenomena, in order to be consistent, must contain in their surface expansion at least the hexadecapole terms.

The barrier penetrability for alpha decay of odd-$A$ nuclei can be calculated in exactly the same way as for even-even nuclei and with much the same results. That such a result is consistent with our physical ideas of barrier penetration is clear enough since we should expect that the properties of the barrier would depend little, if at all, upon the presence of an extra nucleon. However, two essentially different types of odd-$A$

TABLE 7–1

The Empirical $b^x_{61}$ and $b^x_{81}$ for Even-Even Alpha Emitters Calculated from the Experimental Hindrance Factors. In Part (a) the Barrier Is Presumed Asymmetric and the Alpha Wave Function at the Nuclear Surface Is Expressed as a Sum of Second-Order Spherical Harmonics. The Expansion Parameters Being $\varepsilon_{2\mu}$. The Values of the Fitted Electric-Quadrupole Tensor Terms $Q_0$ and $Q_{\pm 2}$ Are also Given. The Quantity $b^t_{81}$ Is Calculated from Eq. (16). In Part (b) the Barrier and Nuclear Surface Is Presumed Symmetric and the Expansion Parameters, $\beta_\lambda$, Are Given. For $Pu^{238}$ and $Cm^{242}$ It Is Estimated that $\beta_6 \approx \beta_8 \approx 0.02$. The Values $b^x_{61}$ and $b^x_{81}$ Are also Given. Part (a) Is from Rafiqullah [138] Part (b) from Froman [94]

(a)

| Parent Nucleus | $b^x_{61}$ | $b^x_{81}$ | $b^t_{81}$ | $\beta_2$ | $\varepsilon_{20}$ | $\varepsilon_{2\pm 2}$ | $Q_0$ | $Q_{\pm 2}$ |
|---|---|---|---|---|---|---|---|---|
| ${}_{90}Th^{230}$ | 0.10 | — | 0.007 | 0.21 | 4.73 | −16.88 | 10.31 | 2.71 |
| ${}_{92}U^{232}$ | 0.31 | — | 0.11 | 0.24 | −2.40 | −72.81 | 12.66 | 1.53 |
| ${}_{94}Pu^{238}$ | 0.30 | 0.18 | 0.10 | 0.24 | −1.62 | −23.58 | 12.65 | 2.08 |
| ${}_{96}Cm^{242}$ | 0.36 | 0.28 | 0.10 | 0.25 | −2.42 | −23.08 | 14.20 | 1.44 |
| ${}_{98}Cf^{246}$ | 0.32 | — | 0.09 | 0.25 | −5.15 | −219.1 | 14.02 | 2.55 |

(b)

| Parent Nucleus | $b^x_{61}$ | $b^x_{81}$ | $\beta_2$ | $\beta_4$ |
|---|---|---|---|---|
| ${}_{90}Th^{230}$ | 0.39 | 0.14 | 0.26 | 0.041 |
| ${}_{92}U^{232}$ | 0.32 | 0.11 | 0.26 | 0.029 |
| ${}_{94}Pu^{238}$ | 0.14 | 0.02 | 0.26 | −0.024 |
| ${}_{96}Cm^{242}$ | 0.06 | 0.00 | 0.26 | −0.041 |
| ${}_{98}Cf^{246}$ | 0.03 | 0.01 | 0.18 | 0.000 |

alpha decays have long been known and are separated simply by their hindrance factors. The favored transitions have hindrance factors for the ground state to ground state decays that are essentially equal to unity. That is, like their even-even neighbors their decays are described very well by the Geiger–Nuttall rule [Eq. (2)]. For the hindered decays the transition rates can be as much as several orders of magnitude slower than is required by Eq. (2). Indeed for such nuclei with hindered ground-to-ground transitions there is often a ground to some excited state transition that is itself unhindered or nearly so. It is now known that the unhindered transitions are characterized by the selection rule $\Delta K = 0$, there being no change of parity. That is, in the transition, the Nilsson orbital of the odd particle remains unchanged. The hindered transitions then result in the change of the Nilsson orbital of the odd particle in going from parent to daughter nucleus. It is quite easy to see, in a qualitative way, why the orbit changing transitions are so hindered over the others. For the unhindered decays the alpha particle can be formed in exactly the same manner and out of the same states as in a neighboring even-even nucleus. The odd particle will make no contribution, for in order to be assimilated into the alpha particle a core pair must be broken. In the hindered transitions, on the other hand, the extra particle must be included into the alpha particle and a specific Nilsson pair must be broken so that the remaining member has the orbital appropriate to the daughter. Clearly this can happen in but one way, which explains the great hindrance. The qualitative features are reproduced quantitatively in the microscopic calculations of the reduced width and the reader should have recourse to the recent literature for a discussion of their details [32].

Alpha decay to the beta and gamma vibrational bands, both in even and in odd-$A$ nuclei, in which they occur, will possess hindrance factors which, in principle, should be straightforward to calculate. For the decay to beta vibrational levels, which have been observed, the hindrance factor must depend mainly on how the penetration of the barrier by the alpha particle is effected by the vibrations. That is, we would simply solve the extremal problem (14) using for the expansion parameters those that are defined in Eq. (13) but at the same time explicitly including the vibrational degree of freedom. This will give rise to integrals similar in form to those in discussion in Chapter 6, Section B on the $E0$ and $E2$ transitions in even-even nuclei. Clearly, no change should be induced in the form or even in the magnitude of the reduced widths since one expects that no great changes in particle configuration arise when nuclear systems are in a beta vibrational mode. On the other hand, effects such as these should be predominant in the decay to gamma vibrational states since a considerable rearrangement of the particle configurations must arise in order to recouple the core to a predominately $K = 2$ band. Of course, this vibration will also affect the penetration of the barrier. However, we should expect the alpha decay to a vibrational state to be much less hindered if the vibration is a beta vibration than if it is a gamma vibration. In the former case, the reduced width will be essentially the same as for the ground state transition. For the latter case, it will be much different and indeed the reduced width will be such that alpha decay to gamma vibrational states will be far more hindered in odd-$A$ nuclei than hindered alpha decays involving a change in the odd-particle configuration.

## B. Beta Decay

The discussion of beta decay within the context of these nuclear models is much facilitated by the observation that the beta-decay operators can be cast into the form of spherical tensors [142]. In what follows we shall first use this property to indicate certain general features of beta decay in deformed nuclei and then outline some of the more detailed calculations of these transitions.

If we call the rank of the operator inducing the beta transition $\lambda$, then from this follows the obvious $K$ selection rule

$$\lambda \geq | K_f - K_i | , \tag{18}$$

which is in addition to the usual angular-momentum selection rule

$$| I_i - I_f | \leq \lambda \leq I_i + I_f . \tag{19}$$

These are identical to similar selection rules for gamma decay. The selection rule (19) is absolute whereas that of (18) depends for its validity on the $K$ purity of the states in question. If $K$ were a good quantum number for states in deformed nuclei, then (18) would be absolute; but since they are not, this selection rule only acts to hinder the given transitions. The degree of $K$-forbiddenness $\nu$ is defined simply as

$$\nu = \Delta K - \lambda$$

and as in gamma transitions the larger $\nu$, the more slowly does the transition proceed over what might otherwise be expected. An example of a very $K$-hindered transition is that from the $7-$ ground state of the odd-odd nucleus $Lu^{176}$ to the $6+$ excited state of $Hf^{176}$. The usual beta-decay selection rules [22] would make this a first forbidden transition with a log ft value of the order of about $7.5 \pm 1.5$. However, the initial state is the head of a $K = 7$ band while the final state belongs to the ground state rotational band with $K = 0$. Thus the transition should be strongly $K$-forbidden with $\nu = 6$. The transition is strongly hindered with log ft = 18.7.

Furthermore, with odd-$A$ nuclei an additional set of selection rules, the asymptotic selection rules, apply which if violated tend to hinder rather than prohibit the transition. In Table 7-2 we give these rules

TABLE 7–2

The Asymptotic Selection Rules for Allowed and First Forbidden Beta Transitions. These Are to be Compared With Similar Gamma Transition Selection Rules of Table 6-1. The Notation in Both Tables Is the Same.

| Matrix element | $\Delta I$ | $\Delta\pi$ | $\Delta K$ | $\Delta N$ | $\Delta n_z$ | $\Delta\Omega$ |
|---|---|---|---|---|---|---|
| $\langle 1 \rangle$ | 0 | no | 0 | 0 | 0 | 0 |
| $\langle \boldsymbol{\sigma} \rangle$ | 0,1 | no | 0, 1 | 0 | 0 | 0 |
| | no $0 \to 0$ | | | | | |
| $\langle \mathbf{r} \rangle$ | 0,1 | yes | 0 | $\pm 1$ | $\pm 1$ | 0 |
| | no $0 \to 0$ | | | | | |
| | | | 1 | 1 | 0 | 1 |
| $\langle \boldsymbol{\alpha} \rangle$ | 0,1 | yes | 0 | $\pm 1$ | $\pm 1$ | 0 |
| | no $0 \to 0$ | | | | | |
| | | | 1 | 1 | 0 | 1 |
| $\langle \gamma_5 \rangle$ | 0 | yes | 0 | $\pm 1$ | $\pm 1$ | 0 |
| | | | | 1 | 0 | 1 |
| $T_0(1, \boldsymbol{\sigma})$ | 0 | yes | 0 | $\pm 1$ | $\pm 1$ | 0 |
| | | | | 1 | 0 | 1 |
| $T_1(1, \boldsymbol{\sigma})$ | 0,1 | yes | 0 | 1 | 0 | 1 |
| | no $0 \to 0$ | | | | | |
| | or $\frac{1}{2} \to \frac{1}{2}$ | | 1 | $\pm 1$ | $\pm 1$ | 0 |
| | | | 1 | 1 | 0 | 1 |
| $T_2(1, \boldsymbol{\sigma})$ | 0, 1, 2 | yes | 0 | 1 | 0 | 1 |
| | no $0 \to 0$ | | | $\pm 1$ | $\pm 1$ | 0 |
| | $\frac{1}{2} \to \frac{1}{2}$ | | | | | |
| | or $1 \to 0$ | | | | | |
| | | | 1 | 1 | 0 | 1 |
| | | | | $\pm 1$ | $\pm 1$ | 0 |
| | | | 2 | 1 | 0 | 1 |

for allowed and first forbidden transitions. They are very similar to the set of rules for odd-$A$ gamma transitions given in Table 6-1. The matrix-element notation follows that of Preston [14].

For allowed and first forbidden transitions we can group the measured ft values not only according to the degree of forbiddenness $\lambda$ but also depending upon whether or not they are hindered. In Table 7-3 we list the limits for these four classes of transitions in deformed nuclei.

TABLE 7-3

Empirical Limits for the Four Classes of Beta Transitions: Allowed Unhindered, Allowed Hindered, First Forbidden Unhindered, and First Forbidden Hindered. The Limits Designated in Columns Three and Four Are the $a$ and $b$, Respectively, of the Relation $a \leq \log ft \leq b$.

| Degree of forbiddenness | Hindrances | Limits of $ft$ values lower | upper |
|---|---|---|---|
| Allowed | $u$ | 4.5 | 5.0 |
| Allowed | $h$ | 6.0 | 7.5 |
| First | $u$ | 5.0 | 7.5 |
| First | $h$ | 7.5 | 8.5 |

Another useful relation involves the intensities of transitions from some initial state in the parent nucleus to two different levels that belong to the same intrinsic state in the daughter. This relation flows from the tensor nature of the interaction operator and is analogous to similar gamma intensity ratios (cf. Eq. (6-56) *et seq.*). This is

$$\frac{f_\lambda t(I_i K_i \to I_{1f} K_f)}{f_\lambda t(I_i K_i \to I_{2f} K_f)} = \left[\frac{C(I_i \lambda I_{1f};\ K_i,\ K_f - K_i,\ K_f)}{C(I_i \lambda I_{2f};\ K_i,\ K_f - K_i,\ K_f)}\right]^2.$$

More detailed investigations of beta transitions require calculating the matrix elements of the various transition operators between appropriate Nilsson levels of parent and daughter. For allowed transitions, for instance, the $f_0 t$ value is

$$f_0 t = 2\pi^3 \ln 2[\langle L_0 \rangle\,(g_V{}^2\,|\mathcal{M}_f|^2 + g_A{}^2\,|\mathcal{M}_\sigma|^2)]^{-1}, \tag{20}$$

where $\langle L_0 \rangle$ is a number near unity and $\mathcal{M}_f$ and $\mathcal{M}_\sigma$ are the Fermi and

Gamow-Teller operators.The simplest class of transitions to consider are the forbidden transitions for which $n = \Delta I - 1$ and $\pi_i \cdot \pi_f = (-1)^{\Delta I+1}$. The $f_n t$ product then depends only on the matrix elements of the $\mathscr{M}_\sigma$ operator that can be expressed as

$$\mathscr{M}_\sigma = \varkappa \sum_j \mathbf{s}_j \cdot \boldsymbol{\nabla}_j [r_j^{n+1} Y_{n+1,\mu}(\theta_j , \varphi_j)] \tau_\pm^j \, . \tag{21}$$

$\varkappa$ is a constant and the only new operators are the isobaric spin stepping operators $\tau_\pm{}^j$. The formal similarity between Eq. (21) and Eq. (6-42b) for the magnetic transitions operator is obvious. Thus for these kinds of transitions the calculations of the $f_n t$ products is done by taking over the results for the magnetic transitions of order $\lambda = n + 1$ and setting $g_s = 1$, $g_l = g_c = 0$ and $\varkappa = e\hbar/2mc$.

Beta transitions between even nuclei are quite similar to gamma transitions in the odd-odd nuclei that were discussed in Chapter 6, Section E. Since such transitions are always between even-even and odd-odd nuclei, we need only to determine the form of the even-even state functions. In order that the matrix elements do not vanish we must single out from the core a pair of like particles, their nature depending upon whether a neutron or a proton is the transforming particle. This like pair is coupled off to $K = 0$ and we obtain a wave function similar to the neutron-proton function that was given in Eq. (5-12). The properly symmetrized and normalized state function is

$$\begin{aligned} | EIMK = 0\rangle &= \sqrt{\frac{2I+1}{32\pi^2}} \sum_{j_p} C^*_{j_p \Omega_p} C_{j_p - \Omega_p} \\ &\times [(\chi^{(1)}_{j_p \Omega_p} \chi^{(2)}_{j_p - \Omega_p} - \chi^{(1)}_{j_p - \Omega_p} \chi^{(2)}_{j_p \Omega_p}) \\ &+ (-1)^{I-2j_p} (\chi^{(1)}_{j_p - \Omega_p} \chi^{(2)}_{j_p \Omega_p} - \chi^{(1)}_{j_p \Omega_p} \chi^{(2)}_{j_p - \Omega_p})] \, D^{I*}_{M0} . \end{aligned} \tag{22}$$

We are specifically assuming the ground state band of a symmetric system (this is a restriction that is not too difficult to generalize) and it can be noted in Eq. (22) that the state function vanishes for $I$ odd since the second bracket is the negative of the first. The subscribt $p$ in Eq. (22) refers to the type of particle undergoing the transition.

In order to calculate the Gamow-Teller matrix elements we use for the other state function that of the odd-odd system that was defined in

Eq. (5-12) and get for the reduced matrix elements of the transition [97]†.

$$D_{GT}(n, I_i \rightarrow I_f) = | C(I_i, n+1, I_f; K_i, K_f - K_i, K_f)$$
$$\times \langle N_{p_f} \Omega^{(1)}_{p_f} | N_{p_i} \Omega^{(1)}_{p_i} \rangle \langle \Omega^{(2)}_p | \mathscr{M}_{GT} (n+1, K_f - K_i) | \Omega^{(2)}_n \rangle |^2. \quad (23)$$

Comparing this equation with Eq. (6-61) for a similar gamma-ray transition, we again observe the product of two integrals. The first is the overlap integral that vanishes unless the orbital of the nontransforming particle is the same in parent and daughter. This again gives rise to so-called "nonoverlap-forbiddenness." The second integral, or those proportional to it, has been calculated in Chapter 6, Section D.

† This is defined in Nilsson [134] in analogy to the similar quantity that is used extensively in electromagnetic transitions as

$$D(n, I_i \rightarrow I_f) = \sum_{\mu M_i} | \langle I_i M_i K_i \Omega_i | \mathscr{M} (n, \mu) | I_f M_f K_f \Omega_f \rangle |^2.$$

CHAPTER 8

# The Nuclear Photoeffect

In this chapter and the next we shall discuss two particular types of specific nuclear interactions which have, or certainly should have, an important place in the investigation of collective nuclear phenomena. It is worthwhile mentioning that both have recently become much more useful as nuclear probes due to new methods of generating or detecting high energy x rays. In one sense, the subject for discussion here is the same as that treated in Chapter 6 in that we shall be mainly concerned with the interaction of photons, or electromagnetic radiation, with nuclear matter. Indeed the photonuclear effect is superficially similar to coulomb excitation, discussed before, in that the initial state involved is invariably the ground state. Here the resemblance ends. Whereas coulomb excitation is essentially an $E2$ absorption process, here we shall be mainly concerned with electric dipole absorption. Furthermore, the nuclear photoeffect is usually considered to be a high energy process in the sense that the energy of the incident photons is greater than the binding energy of the last nucleon. Thus our interest will focus on processes such as the $(\gamma, n)$ reaction although we shall say a few words later about inelastic photon scattering.

Generally, photonuclear physics is divided into two broad categories depending upon the type of particle inducing the reaction. On the one hand, there are photon induced reactions, the branch of the subject of most immediate concern. Electron induced reactions, on the other hand, although not as yet as well developed a branch of photophysics as photon induced reactions, will become relatively more important as more high energy, high intensity electron accelerators become available. Indeed they will probably become the most important tool to investigate collective phenomena whose energy is above the photon, giant dipole resonance. We shall mention later a few inelastic electron scattering experiments but one should refer to recent reviews for a more detailed treatment [24, 29].

In Chapter 6, Section A we explained the large hindrance factors observed in low energy $E1$ transitions as being due to the coincidence of the centers of mass and charge in collective models of low lying nuclear structure. The observed $E1$ transitions involve relatively few nucleons of the system. Clearly, then, in order to obtain the large amount of dipole absorption observed in the giant resonance we must assume processes in which the centers of mass and charge suffer a relative displacement. In heavy nuclei ($A \gtrsim 50$) the most successful models are hydrodynamic ones in which such a relative displacement occurs. The first theoretical discussion of this process [103] put forward three essentially different mechanisms giving such relative displacements. These were:

1. That each proton executes harmonic oscillations about its equilibrium position. The restoring force is the same for all protons.
2. The nucleus consists of a proton fluid and neutron fluid executing density fluctuations within a fixed envelope. The restoring forces per unit mass were assumed proportional to the local density gradient.
3. The dipole motion of neutrons and protons is that of hard interpenetrating spheres; the total restoring force was taken to be proportional to the nuclear surface.

If $E_m$ is the energy at which the cross section is maximum, then for the first model it is constant for all nuclei; for the two-fluid model it is proportional to $1/R$ or $A^{-1/3}$; whereas in the hard-sphere case it is proportional to $1/\sqrt{R}$ or $A^{-1/6}$. The authors [103] felt that the available data supported the third model which they developed somewhat further. Since these suggestions were put forward very soon after the first announcement of the giant resonance effect [46] experimental evidence to support a choice of model was sparse; however, more recent discussions [42] indicate $E_m$ more nearly proportional to $A^{-1/3}$ so that the current choice is with the second model and its adaptions and variations [27]. We shall consider these in some detail, beginning with a consideration of the classical motion of a two-fluid system which was initially applied to this nuclear problem by Steinwedel et al. [149]. They assumed spherical symmetry; however, Danos later extended these calculations to axially deformed systems [67] and Inopin and Okamoto, to asymmetrically deformed systems [110]. We shall use the techniques of classical hydrodynamics [38] starting from a variational principle.

Hamilton's principle for a continuous, classical system takes the form

$$\delta \int L\,dt = \delta \int dt \int_v \mathscr{L}\,d\tau\,, \tag{1}$$

where $L$ and $\mathscr{L}$ are, respectively, the Lagrangian and the Lagrange density this being the difference of kinetic and potential energy densities. The first of these latter is easily written down as the logical extension of Eq. (2-5) to our two fluid problem

$$\mathscr{T} = \frac{1}{2}\,(\varrho_p\,\mathbf{v}_p{}^2 + \varrho_n\,\mathbf{v}_n{}^2) \tag{2}$$

where $\varrho_i$, $\mathbf{v}_i$ are the mass densities and velocities of the proton and neutron fluids. In what follows, we shall assume that the mass and charge densities are proportional, that is, $\varrho_p = \gamma\varrho_c$ with $\gamma = M/e$, where $M$ is the nucleon mass. The important idea of Steinwedel et al. [149] was that the restoring force here arises from the symmetry energy defined in the semiempirical mass formula as [14]

$$V_S = \varkappa'(N-Z)^2/A;$$

thus the potential energy density is just

$$\mathscr{V} = \varkappa\,\frac{(\varrho_n - \varrho_p)^2}{\varrho_0} \tag{3}$$

where $\varrho_0$ is the sum of the separate densities and $\varkappa = \varkappa'/M$, $M$ the nucleon mass.

Before proceeding we must specify the constraints to which the system is subjected. We assume the total fluid system is incompressible although not the two parts separately; and of course, we require the total neutron and proton mass to be conserved during the absorption process. Thus we have

$$\varrho_p + \varrho_n = \varrho_0\,, \text{ a constant} \tag{4a}$$

$$\dot{\varrho}_p + \boldsymbol{\nabla}\cdot\varrho_p\mathbf{v}_p = 0 \tag{4b}$$

$$\dot{\varrho}_n + \nabla \cdot \varrho_n \mathbf{v}_n = 0 \tag{4c}$$

$$\mathbf{r} \cdot \mathbf{v}_i \,|_{r=R_0} = 0, \qquad i = p,\ n \tag{4d}$$

this latter being the envelope boundary condition. The process is subjected to the Lorentz forces as a driving force

$$\mathbf{F} = q(\mathbf{E} + \mathbf{v}_p \times \mathbf{B}) = q\mathscr{F} \tag{5}$$

in suitable units, and presumably to dissipative forces which give rise to the fairly large measured width of the resonance. Finally, during absorption we must maintain the total number of nucleons constant. This constraint is nonholonomic and must be included through the use of a Lagrange multiplier, $\lambda$ say, by simply adding to the integral of Eq. (1) a term of the form

$$\delta\mathscr{S} = -\lambda\delta\varrho_0\,. \tag{6}$$

The force condition from (5) is easily added by introducing it as a generalized force acting on the proton fluid

$$\delta\mathscr{E} = \gamma\varrho_p\mathscr{F} \cdot \delta\mathbf{r}_p \tag{7}$$

$\mathscr{F}$ being defined in Eq. (5).

The dissipative forces give rise to the broadening of the resonance by coupling to various, and here unspecified, internal degrees of freedom. They may be included in the variational principle through the dissipation density function

$$\mathscr{D} = -\varepsilon\frac{\varrho_n\varrho_p}{\varrho_0}(\mathbf{v}_n - \mathbf{v}_p)^2 = -\varepsilon\frac{\varrho_n\varrho_p}{\varrho_0}\mathbf{v}^2 \tag{8}$$

which depends on the relative velocity $\mathbf{v}$ and vanishes if either nucleon density is zero. The associated, generalized force density is simply

$$\mathscr{X}_i = \frac{\partial\mathscr{D}}{\partial\dot{r}_i}$$

which is also to be added to Eq. (1).

The variation to be carried out is then just

$$\delta\int_{t_1}^{t_2} L\,dt = \int_{t_1}^{t_2} dt \int_v (\delta\mathscr{T} - \delta\mathscr{V} + \delta\mathscr{S} + \delta\mathscr{E} + \sum_i \mathscr{X}_i \cdot \partial\mathbf{r}_i)\,d\tau \tag{9}$$

which is done in the usual way subject to the boundary conditions (4).

From this we obtain

$$\frac{\partial \mathbf{v}}{\partial t} + \frac{8\varkappa}{\varrho_0} \operatorname{grad} \varrho_p - \mathscr{F} + 2\varepsilon \mathbf{v} + (\mathbf{v}_p \cdot \operatorname{grad}) \mathbf{v}_p - (\mathbf{v}_n \cdot \operatorname{grad}) \mathbf{v}_n = 0 \quad (10)$$

and

$$\frac{\partial \mathbf{V}}{\partial t} - \operatorname{grad} \left[ \varkappa \left( \frac{\varrho_0 + 2\varrho_p}{\varrho_0} \right)^2 - \lambda \right] - \frac{\gamma \varrho_p}{\varrho_0} \mathscr{F}$$
$$- \frac{\mathbf{v}}{\varrho_0} \frac{\partial \varrho_p}{\partial t} + \frac{\varrho_p}{\varrho_0} (\mathbf{v}_p \cdot \operatorname{grad}) \mathbf{v}_p + \frac{\varrho_n}{\varrho_0} (\mathbf{v}_n \cdot \operatorname{grad}) \mathbf{v}_n = 0 \quad (11)$$

which are exact. The quantity $\mathbf{V}$ is the center of mass velocity.

In order to obtain a wave equation we linearize Eqs. (10) and (11), take the divergence of the former, and so obtain the wave equation with damping (setting $2\varepsilon = \Gamma$, the resonance width)

$$\left[ \frac{8\varkappa}{\varrho_0{}^2} \nabla^2 - \frac{\Gamma}{\varrho_p \varrho_n} \frac{\partial}{\partial t} - \frac{1}{\varrho_p \varrho_n} \frac{\partial^2}{\partial t^2} \right] \varrho_p = 0. \quad (12)$$

The linearized form of Eq. (11) does not yield a wave equation—it leads only to classical Thompson scattering. If we assume a harmonic time dependence for the solutions of (12) of the form

$$\varrho_p(\mathbf{r}, t) = \varrho_p(0)[1 + \eta(\mathbf{r}) \varrho^{-i\omega t}], \quad (13)$$

the wave equation becomes the scalar Helmholtz equation

$$(\nabla^2 + k^2) \eta(\mathbf{r}) = 0, \quad (14)$$

where the wave number is

$$k^2 = \omega^2 \left( 1 + i \frac{\Gamma}{\omega} \right) (A^2/8\, \varkappa NZ) \quad (15)$$

and Eq. (14) is subject to the boundary conditions (4). This system is identical with that arising from the problem of the acoustical vibrations of a gas confined in a rigid spherical shell. As might be expected, the solutions of this latter problem were discussed quite sometime ago by Rayleigh [17] and the eigenvalue problem may be written

$$\tan kr = 2kr/(2 - k^2 r^2). \quad (16a)$$

The lowest nontrivial solution was given by Rayleigh as

$$(kr)_1 = 119.26\,\pi/180 = 2.08. \tag{16b}$$

He gave some of the higher modes also. These, however, are of no concern in the nuclear problem since the damping is such that no higher resonance has ever been seen. In any event, this lowest mode leads to the energy

$$E_m = \hbar\omega_1 = \alpha A^{-1/3}\ \text{MeV}, \tag{16c}$$

where $\alpha$ is simply evaluated from the foregoing equations of motion

$$\alpha = \frac{2.08\,\hbar}{r_0}\sqrt{\frac{8\varkappa'}{M}\,\frac{NZ}{A^2}} \cong 70\ \text{MeV},$$

which is quite close to the experimental value of 80 MeV.

Another point worth investigating is the integrated cross section that should not violate the sum rule [12]. If $I_0$ be the incident photon-flux then, the absorption cross section is just

$$\sigma = \frac{\langle \mathbf{E} \cdot \mathbf{P} \rangle}{I_0},$$

where $\mathbf{P}$ is the polarization again in some appropriate unit system.† The integrated cross section is then calculated in a straightforward manner once the dipole moment is calculated from our solutions in the foregoing. This yields

$$\int_0^\infty \sigma\, d\omega = \frac{2\pi^2 e^2}{Mc}\left[\frac{NZ}{A}\sum_{n=1}^{\infty}\frac{2}{(kr)_n^2 - 2}\right],$$

where the quantity in brackets is the summed oscillator strength. The sum itself is unity, thus the sum rule is fulfilled; however, from Eq. (16b) we can show that the lowest mode accounts for over 85% of the

† The question of the appropriate units to be used with various electromagnetic quantities appearing particularly in this chapter is left open. Therefore, all such quantities as $c$, $\varepsilon_0$, $\mu_0$, and so on, have been deleted and may easily be supplied by the reader once he has selected his unit system following the dictates of taste, current fashion, or perhaps some rationalized method.

integrated cross section. Thus higher resonances, if they exist, could not be seen by current methods.

The model is indeed quite simple to give this close agreement with experiment. Several obvious physical effects have been neglected. In the first place, coulomb repulsion of the charged fluid will alter the restoring forces [assumed due only to the symmetry energy (3)]. Exchange forces have long been known to play a role in the photoeffect and have also been ignored. Finally Eq. (4d) implies a rigid and spherical surface. The high energy electron scattering experiments are consistent with a softer surface with a "skin thickness" wherein the charge density falls from 90 to 10% of its maximum value [see Eq. (9-1)]. These effects are considered more fully elsewhere [12].

The lack of sphericity of the nuclear surface is easier to deal with and certainly germane to the discussion here. A classical fluid with a rigid, but spheroidal, boundary would "ring" with two fundamental modes. The wavelengths of each would be proportional to the length of the semiaxes. Thus photon absorption in deformed nuclei should display a double (at least) humped cross section curve. The energy of each peak should be in a reciprocal ratio to the ratio of the semiaxes. Furthermore, the mode normal to the symmetry axis will be doubly degenerate hence one hump of the cross section curve will have an area twice the second. Since the peak energies are related to the eccentricity their splitting will be a measure of the (ground state) quadrupole moment. This then offers the possibility of an independent measurement of this moment.

This actual situation is not quite as simple as depicted above since the fundamental frequencies are somewhat shape dependent. Danos has considered the appropriate boundary value problem in some detail [67] solving the scalar Helmholtz equation (14) with a spheroidal boundary condition for the $l = 1$, $m = \pm 1$ case which together with MacLaurin's $l = 1 m = 0$ solution of a similar acoustical problem [126] led Danos to conclude that the relation between the fundamental frequencies is quite adequately given by

$$\frac{\omega_b}{\omega_a} = 0.911(a/b) + 0.089. \tag{17}$$

Here $a$ is along the symmetry axis.

Most experiments are done with unoriented targets so that the nuclear symmetry axis must be averaged over all directions. This gives for the average cross section

$$\langle\sigma(E)\rangle = \frac{1}{3}\,\sigma_a(E) + \frac{2}{3}\,\sigma_b(E).$$

Finally, if the widths $\Gamma_i$ $(i = a, b)$ are functions of the resonance energies then

$$\frac{(\sigma_a)_{\max}}{(\sigma_b)_{\max}} = \frac{\Gamma_b}{2\Gamma_a}.$$

These predictions were first confirmed by Fuller and Weiss [95] in a betatron experiment with tantalum and terbium targets. More recently, experiments using monoenergetic photon beams with various targets of deformed nuclei have obtained values of the quadrupole moments consistent with those obtained by other means (notably by coulomb excitation).

In Chapter 3 we discussed nuclear models with an ellipsoidal shape and whereas these systems are geometrically not very asymmetric, the fact that they have three unequal semiaxes should in principal give a triple humped absorption curve. Both Inopin and Okamoto have considered this phase of the problem in some detail using the different approaches and getting different results although their general predictions are similar [110]. This matter has been reviewed in some detail elsewhere [28] and will not be repeated here except to note that studies with monoenergetic photon beams using deformed targets do not support an ellipsoidal shape [53]. Model fitting procedures indicate that the even-even platinum isotopes should probably possess the maximum asymmetry ($\gamma = 30^{\circ}$) Recent betatron experiments using the enriched isotopes of platinum failed to show a triple humped absorption curve [107]. Even though the data reduction techniques of betatron experiments throw some doubt upon the details of the cross section curves, especially above the principal maximum, there would seem to be little doubt that no deformed nuclei are very asymmetric in shape. These nuclei are on the upper edge of their deformed region being often referred to as transitional or vibrational nuclei. The fact that they possess large amplitude beta (and possibly gamma) vibrations probably insures that any dipole fine structure is averaged out. This question of vibrations plays an important role and we shall return to it shortly.

An interesting manifestation of nuclear deformation arises when one sets out to determine the photon scattering cross section. If the scattering amplitude $f(\theta)$ is simply a scalar function then there exists a unique relation between the total photon absorption cross section and the differential scattering cross section. The differential cross section is just

$$\sigma_s(\theta) = |f(\theta)|^2 \tag{18}$$

whereas the optical theorem yields for the total absorption cross section [16]

$$\sigma_a = \frac{4\pi}{k} Im[f(0)]. \tag{19}$$

Thus for $E1$ scattering the scattering cross section should behave as $(1 + \cos^2\theta)$. For deformed nuclei it has been shown that there occurs an additional term in the scattering cross section of the form $(13 + \cos^2\theta)$ which is related to the nuclear polarizability [96]. The differential scattering is then said to consist of two terms, a scalar part [Eq. 18] and a tensor part

$$\sigma(\theta) = \sigma_s(\theta) + \sigma_t(\theta). \tag{20}$$

If we call the dipole scattering amplitudes $f_K$, then the terms of Eq. (20) can be put into the form [27]

$$\begin{aligned} \sigma_s(\theta) &= \left| \frac{f_0 + 2f_1}{3} - \frac{Ze^2}{AMc^2} \right|^2 \frac{(1+\cos^2\theta)}{2} \\ \sigma_t(\theta) &= |f_0 - f_1|^2 \frac{(13+\cos^2\theta)}{90}. \end{aligned} \tag{21}$$

The expression for the absorption cross section being the sum of two terms, will also be altered. If we call $F_2$ the second-order orientation parameter [153] then for an unpolarized photon beam incident on a deformed target oriented at an angle $\theta$ to the beam

$$\sigma_a = \sigma_{as} + \sigma_{at} F_2 P_2(\cos\theta).$$

Here

$$\sigma_{as} = \frac{4\pi}{k} Im\left(\frac{f_0 + 2f_1}{3}\right)$$

and

$$\sigma_{at} = -\frac{4\pi}{k} \frac{I^2}{(I+1)(2I+3)} Im(f_0 - f_1),$$

which vanishes for even-even nuclei. The relative size of the two terms is model dependent and has been calculated for simple hydrodynamic models [96] as well as for more complicated models with various interactions [44].

We now turn to a discussion of these more detailed models. Heretofore we have considered only a static interaction with the electromagnetic field in that the nucleus has been taken as only a rigid, although deformed, system. In order to see what role the collective nuclear coordinates play in the interaction it is perhaps best to start with the spherical vibrator model. The surface expansion of Eq. (2-1) will now include the three $\lambda = 1$ terms. In order to construct an appropriate interaction term for the Hamiltonian we need only observe the simple rules that it be a spherical tensor of rank zero and be invariant under time reversal (hence have an even number of time derivatives) and point reflection. We choose an operator with no time derivatives and meet the other requirements by noting that if we couple the dipole part of the radiation field with dipole oscillations of the nuclear fluid then in lowest order in the $\alpha_{\lambda\mu}$ operators these will just couple to the quadrupole mode. The operator is then

$$H_{\text{int}} = \mathscr{K} \sum_{\mu,\nu} (-1)^{\mu} C(112; \nu, \mu - \nu, \mu) \alpha_{1\nu} \alpha_{1\mu-\nu} \alpha_{2-\mu}. \qquad (22)$$

The calculation is most easily carried out in the second quantized formalism making use of the creation and destruction operators defined in Eq. (2-13). The basis is now an uncoupled product basis of the form

$$| \lambda = 1, N, L, \mu\rangle \, | \lambda = 2, \mathscr{N}, \mathscr{L}, \nu\rangle.$$

An investigation such as this has been carried out [122] with up to 16 surfons yielding a very rich spectrum. Its usefulness, except as a theoretical model, is somewhat questionable since the pure surfon model does not reproduce many important details of lower energy structure even in the transition regions. Furthermore, the question of the relative importance of higher order terms coupling to several different $\lambda$ modes with fewer surfons vis-à-vis the single term of Eq. (22) with such a large number of vibrational quanta has not been considered. Indeed such high energy transitions as those involved in the giant dipole resonance must surely couple to more than just a single $\lambda$ mode of the system. High energy electron scattering data indicate quite strong coupling to $\lambda = 4$ and 5 modes in heavy nuclei [113].

In order to apply this type of interaction to models of deformed nuclei, either one may start by transforming the total Hamiltonian for dipole and quadrupole oscillations with the interaction term (22) from laboratory to body-coordinate systems, or one may start by writing down a similar operator in the body system [68]. As usual we consider first deformed even-even nuclei and couple the dipole oscillations to the low lying rotational and vibrational structure. Furthermore, we restrict the discussion to the adiabatic approximation (Chapter 3, Section B) since the characteristic energies involved differ one from another by about an order of magnitude. That is $E_{\text{rot}} \ll E_{\text{vib}} \ll E_{\text{dipole}}$. The problem then is quite analogous to that of odd-$A$ deformed nuclei (Chapter 4), since if $\mathbf{j}_d$ is the angular momentum operator for the dipole degrees of freedom then the total angular momentum is

$$\mathbf{I} = \mathbf{L} + \mathbf{j}_d \tag{23}$$

which is formally identical with Eq. (4-1).

The Hamiltonian for the system is then

$$H = H_R + H_{\text{vib}} + H_d + H_{v-R} + H_{v-d} + H_{d-R} \tag{24}$$

to which we should add a term due to the radiation field. Most of these terms have been discussed before with their associated Schrödinger equations. The wave equation for $H_{vib}$ involves beta and gamma vibrations and upon separation is just that of Eqs. (3-27) and (3-28). (The treatment of Danos and Greiner [68] does not use the $\beta$ and $\gamma$ coordinates but the body expansion parameters $a_{2\mu}$ of Eq. (2-2). A critique of this model appears elsewhere; however, the calculation of interest here may proceed from any formulation of the quadrupole rotation-vibration problem.) The term $H_{v-R}$ is that of Eq. (2-32) whereas $H_{v-d}$ is identical with the operator defined in Eq. (4-46). If one solves the dipole problem in the second quantized representation, then both $H_d$ and $H_{v-d}$ have a simple representation

$$H_d + H_{v-d} = \sum_{\varkappa} \hbar\omega_\varkappa b_\varkappa^+ b_\varkappa ,$$

which represents the dipole and dipole-quadrupole vibration interaction.

The basis to be used to solve the complete equation is the product of vibrational, rotational, and dipole terms

$$\psi = \varphi_{n_\beta n_\gamma} D^{I*}_{MK} \eta_s \tag{25}$$

in that order. The quantum numbers $n_\beta$ and $n_\gamma$ in this equation have been defined in Eqs. (3-29b) and (3-30b), respectively. The dipole mode is given by $s = 0$ for excitations parallel to the symmetry axis and by $s = \pm 1$ for excitations normal to the symmetry axis. As usual the function $\psi$ must be symmetrized with respect to the operators $T_1$ and $T_2{}^2$ defined in Chapter 3, Section A. The Schrödinger equation for the system formed from Eqs. (24) and (25) has for solutions

$$\begin{aligned} E(I, K, n_\beta, n_\gamma, s) \\ &= [I(I+1) - K^2 - |s|]E_R + \hbar\omega_\beta(n_\beta + \tfrac{1}{2}) \\ &\quad + \hbar\omega_\gamma(n_\gamma + 1) + \hbar\omega_s . \end{aligned} \tag{26}$$

In this equation the energies $E_R$, $\hbar\omega_\beta$, and $\hbar\omega_\gamma$ are to be considered as parameters that are determined from the low lying level structure of the nucleus concerned.

An interesting consequence of unifying the collective quadrupole modes and the giant dipole oscillations is that even by starting with a nuclear system with a symmetry axis the coupling between dipole and quadrupole collective modes stabilizes the figure of the system about an asymmetric shape. This then results in removing the degeneracy of the higher frequency mode of the giant resonance, splitting it in two. Not only that, but the gamma vibrational mode, since it is a spin two oscillation, further splits each resonance and gives it a satellite line that takes some of the dipole strength. Since these splittings are not great and since the dipole strength originally in the upper resonance is distributed over several more lines, the resonance lines so arising are probably not observable by current techniques but contribute to the general breadth of the resonance. Indeed the problem of observing the individual resonances predicted by this model of even-even nuclei is almost insurmountable. No measurement of the giant resonance on isotopically pure even-even deformed targets has ever been reported and the only work reported on enriched even-even targets (of platinum) did not use monoenergetic photon beams.

A better test of the model would be for odd-$A$ nuclei since many are monoisotopic; however, as we saw in Chapter 4 one cannot just take the theory over directly for such systems. Not only does the odd-particle coupling lead to greater complexity of the level structure but also the spectrum will be greatly enriched by the fact that the ground state spin will not be zero. Thus the dipole modes do not couple simply to 1 − states but to a group of three states of spin $I_g$, $I_{g\pm1}$. A unified odd-$A$ model has been studied in some detail [70] by adding to Eq. (24) the odd particle Hamiltonian defined in Eq. (4-14) with a term to account for interaction between core vibrations and the particle degrees of freedom. In this investigation the simplifying assumption of neglecting $H_{v-R}$ was made. The analysis is quite complex but straightforward yielding the expected, but simple, result that the odd particle does not alter the main features of the giant resonance. The details of the resonance, that is the individual lines composing it, are much richer than for the even-even case. Present experimental techniques are such, even when using monoenergetic photon beams, that this model is adequately tested in odd-$A$ monoisotopes by fitting to the expression (26). For instance by making use of the low energy structure of $Er^{166}$ to obtain $E_R$, $\hbar\omega_\beta$ and $\hbar\omega_\gamma$ it has been possible to obtain very good agreement with the measured $(\gamma, n)$ spectrum of $Ho^{165}$. In this work [69] the resonance widths $\Gamma_k$ were fit to an expression of the form

$$\Gamma_k = \Gamma_0 E_k{}^\delta$$

with the exponent determined to be 2.2 whereas the deformation parameter $\beta_0$ was obtained from the separation of the peaks.

In closing this chapter we shall mention briefly some experiments and calculations within that other branch of photophysics, high energy electron scattering. It has been suggested that because the electron-nucleus interaction is basically an electromagnetic one "... inelastic (electron) scattering can be considered as a kind of coulomb excitation of nuclei" [24]. Because of the very high energy available to be transferred to the nucleus and because at all bombarding energies the interaction remains electromagnetic (as opposed to coulomb excitation with alpha particles, say, in which the specific nuclear part of the scattering increases with bombarding energy) inelastic electron scattering would seem to be an ideal probe of collective nuclear modes.

Inelastic electron scattering with incident energies up to some 600 MeV have excited $\lambda = 2$, 3, and 4 vibrational modes in nondeformed nuclei (for example, Fe, Co, Ni, In, Pb, and Bi) with an enhancement over the single-particle rate of up to 40 times whereas in the heaviest spherical nuclei, $\lambda = 5$ modes have apparently been seen with an enhancement of about a factor of ten. The data are generally analyzed in relation to scattering form factors which relate the calculated Mott differential scattering cross-section [16] to the observed cross section

$$\sigma_{\mathrm{obs}} = \sigma_{\mathrm{M}} \mid F \mid^2 \tag{27}$$

which in Born approximation is simply

$$\mid F \mid^2 = \left| \int e^{i\mathbf{q}\cdot\mathbf{r}} \langle f \mid \mathscr{E}^L \mid i \rangle \, d\tau \right|^2 \tag{28}$$

$\mathbf{q}$ being the momentum transferred while the operator in the matrix element has been defined quite generally in Chapter 6. The Born approximation has been adequate to determine in these nuclei the multipolarity of the induced transition and hence the order of the collective mode excited. Furthermore, this approximation provides useful fits of the form factors and transitions strengths consistent with other methods [113]. One curious result was with the only deformed nucleus studied, $Ta^{181}$. In this nucleus the electrons seem to excite the higher order vibrational modes only weakly if at all. Since this experiment was done with a bulk target, it would seem that more useful information might be obtained from inelastic scattering using aligned deformed nuclear targets. A theoretical investigation of this process has been made using the Schiff–Tiemann approximation and it was found that nuclear deformation significantly alters the differential cross section for the case studied (holmium) [105].

In general such calculations as these form an interesting starting point for further investigation of higher energy collective properties. Unfortunately nuclear alignment is not available as an experimental tool in even-even nuclei and other techniques will have to be developed for them or their structure inferred from their odd-$A$ neighbors.

CHAPTER 9

# Mu-Mesic Atoms and Collective Structure

In this final chapter we shall discuss a method, which while only now being exploited, should become a very powerful and important one for the investigation of collective properties of heavy nuclei. The negative mu-meson because of its large mass and very small specific nuclear interaction can be used as a probe of nuclear structure. This is easily appreciated when it is observed that the muon $k$-shell radius is somewhat less than the nuclear radius for a heavy element such as lead. (By this we mean the quantity $\sqrt{\langle r^2 \rangle}$. The Bohr radius for a mu-mesic atom, $a_0 = \hbar^2/mZe^2$, is much less than the nuclear radius of lead; however, a point nucleus approximation such as this is obviously not valid for mesic atoms.) Thus in mesic lead the gamma ray associated with the electric dipole $2P$–$1S$ transition has some 6 MeV of energy [90]. Because in the initial and final states of such transitions the muon spends a large fraction of time within the nucleus it was postulated quite some time ago that the characteristics of these decay gamma rays should be affected by certain nuclear properties [41, 157]. Such mesic gamma rays were first detected and their properties crudely measured in an expansion cloud chamber by Chang [61] who was able to conclude that some of the observed gamma rays were extranuclear in origin.

The nuclear charge distribution will profoundly effect the characteristics of the inner muon orbits so that by simply measuring the mesic $E1$ transition energies, particularly the $2P$–$1S$ energy, it is possible to determine this charge distribution. Indeed the charge radius was the first nuclear property to be investigated in detail by this mesic probe. In 1953

Fitch and Rainwater [90] measured $2P$–$1S$ transition energies for nine atoms from aluminum to bismuth. The analysis of this data [66, 90] was shown to be consistent with a nuclear radius $R = r_0 A^{1/3}$F with $r_0 = 1.2$ (previously) this radius parameter had been taken as $r_0 = 1.4$). At this same time Wheeler suggested other possible uses for this type of experiment [157]. He pointed out that the muon probe would be ideal to investigate (1) nuclear quadrupole moments, (2) the muon magnetic moment, (3) the details of the nuclear charge distribution, and finally (4) the nuclear polarizability and compressibility. To this we can now add (5) the distribution of nuclear magnetization which can be observed through the Bohr–Weisskopf effect [118]. Of particular interest here is (1) just mentioned and Wheeler pointed out [157] that the $2P_{3/2}$ level would in turn be split if the nucleus in question had a quadrupole moment (and $I \geq 1$). This effect can be called the static electric quadrupole hyperfine splitting and also occurs in ordinary atoms. Clearly, it will be much larger in mesic atoms than in ordinary atoms and for deformed nuclei will be more nearly equal in magnitude to the fine structure splitting. This results principally from the fact that the muon magnetic dipole moment is some two hundred times smaller than that of the electron. For uranium the $2P_{1/2} - 2P_{3/2}$ doublet splitting is about 235 keV [92] whereas the hyperfine splitting is of the order of 10–20 keV. [158].

It was pointed out by Wilets [158] and Jacobsohn [111] that in such mesic atoms the hyperfine effect was in principle different from that in ordinary atoms. Because the $2P$–$1S$ energy difference is of the order of several MeV, which is an order of magnitude or more greater than the lowest nuclear energy level spacing, the muon does not interact with a rigid nucleus in its ground state but with a nucleus in a mixture of several states. Thus the muon in going from the $2P$ to $1S$ atomic levels can leave the nucleus in some excited state (for uranium, the probability of leaving the nucleus in the first rotational excited state has been calculated to be about 50% [158]). This sort of quadrupole interaction can be called the dynamic electric quadrupole hyperfine interaction. Its effects have recently been measured in several deformed nuclei [85] and subsequently we shall review the theory of these hyperfine interactions.

In the following discussion we shall specifically ignore any magnetic hyperfine effects that arise from the magnetic dipole interaction, since the ratio of electric quadrupole to magnetic dipole energy is of the order of

$$\frac{\langle e^2Q/R_0^{\,3}\rangle}{\langle\mu_n\mu_\mu/R_0^{\,3}\rangle}=\frac{e^2Q}{\mu_n\mu_\mu}\approx 10^2.$$

The spin-orbit fine structure interaction will be included in the unperturbed Hamiltonian (and arises naturally if this is the Dirac Hamiltonian). It may be necessary to include magnetic dipole effects for the upper edges of the deformed regions where nuclei have larger dipole moments and smaller quadrupole moments. (Bismuth is an extreme case of this, since it is essentially spherical with a very large magnetic dipole moment of 4.04 nm and a very small electric quadrupole moment of about 0.4 b).

From a theoretical point of view the investigation of the mesic interaction with collective modes of the nucleus is much more clear cut than similar investigations of electron interactions since Gauss' law of classical electromagnetism assures us that for the lower orbits of interest electron screening plays a relatively unimportant role. Quite detailed mesic orbit calculations using the appropriate Dirac–Hamiltonian with a Fermi-type potential of the form

$$\varrho(\mathbf{r})=\varrho_0\left[1+\exp\left(\frac{r-c}{s}\right)\right]^{-1} \tag{1}$$

have been done. The half-radius $c$ and the skin thickness $s$ are either fit to the high energy electron scattering data [31] or to the observed mesic $K$ and $L$ x-ray energies [85]. The normalization of Eq. (1) is simply

$$\int\varrho(\mathbf{r})\,d\mathbf{r}=Ze.$$

It is customary to include in such calculations the effects of vacuum polarization. For some time now the results of such a numerical calculation have been available [92]. In this work cited the $2P$–$1S$ transition energies for 34 elements for atomic number on the range $4\leq Z\leq 92$ are presented as well as numerical tables of the $1S$ state functions for $Z(r,\Delta r)$ equal to 4(375.0, 0.5), 22(117.5, 0.5), 30(75.0, 0.5), 51(63.0, 0.5), 82(25.0, 0.5).

For heavy elements it has been shown that a nonrelativistic treatment of the problem using a uniform charge distribution and treating the spin-orbit interaction by first-order perturbation theory yields eigenvalues within a percent or so of the measured values [54, 162]. In the following discussion we shall use such nonrelativistic assumptions, where necessary, because of the convenience and simplicity deriving therefrom, although recognizing that detailed verification and correlation of predictions of nuclear structure properties with experiment will often require numerical integration of the Dirac equation with more realistic charge distributions.

One consequence of the finite nuclear size that is immediately evident is that the $2S$ state is, unfortunately perhaps, not metastable since it is well above the $2P$ level. This can be seen in the nonrelativistic, uniform charge approximation. The charge uniformity leads to a radial potential function within the nuclear surface which is identical in form to that of a displaced three-dimensional harmonic oscillator [162]. From the well-known structure of the oscillator solutions it is seen that the $2S$ level moves above the $2P$ level, the relative displacement being greater the larger the nucleus (that is, the greater $Z$).

First, let us consider the interaction of a negative muon with an even-even nucleus after which we can consider what effects might arise from an extra-core nucleon or two. The Hamiltonian for the mesic atom can be written as

$$H = H_n + H_\mu + H_e - \sum_p \frac{e^2}{|r_p - r_\mu|} + \sum_e \frac{e^2}{|r_\mu - r_e|} \tag{2}$$

the interaction Hamiltonian being the last two terms and we tacitly assume the atom consists of a single muon and $(Z - 1)$ electrons. The first three terms are associated with the unperturbed problems

$$H_n \varphi_j = E_n \varphi_j \tag{3a}$$

$$H_\mu \chi_i = E_{\mu_i} \chi_i \tag{3b}$$

$$H_e \psi_k = E_{e_k} \psi_k \tag{3c}$$

and are nuclear, mu-mesic, and electronic equations of motion, respectively. In the present discussion we may neglect completely the electronic part of the problem (Eq. 3c) since for the lowest mu orbits Gauss' theorem implies that there will be little electron screening. We will then also neglect the last term in Eq. (2) whose primary contribution is in nonradiative Auger transitions. This is not to say that such transitions are uninteresting, on the contrary. Unlike their atomic analogs, they are allowed in all such atoms and what measurements have been made indicate that they are well explained by theory [148]. However, they would appear to be little influenced by nuclear structure so we shall leave them for another time and place.

The mu-nuclear interaction term is

$$V'_{\mu p} = - \sum_{p=1}^{Z} \frac{e^2}{|\mathbf{r}_p - \mathbf{r}_\mu|}, \tag{4}$$

which is to be expanded in terms of spherical tensors. The coordinates to be used are those of the muon and nuclear protons relative to the laboratory $(r_\mu, \theta_\mu, \varphi_\mu)$, $(r_p, \theta_p, \varphi_p)$ and relative to a nuclear body-fixed system $(r_\mu, \theta_\mu', \varphi_\mu')$, $(r_p', \theta_p', \varphi_p')$ which are related by the usual Euler angles $\theta_i$. Finally, the angle between the muon and nuclear proton radius vectors is denoted by $\theta_{\mu z}$. Thus (4) becomes

$$V_{\mu p} = e^2 \sum_{p=1}^{Z} \sum_{l=0}^{\infty} \frac{r_<^{\,l}}{r_>^{l+1}} P_l(\cos \mu_{\mu Z}) \tag{5}$$

where $r_<$ $(r_>)$ is the lesser (greater) of $r_p$ or $r_\mu$. On calling $V_0$ the spherically symmetric part of this potential function, the interaction Hamiltonian is just

$$H' = V - V_0,$$

$V_0$ being the central part of the potential in the problem represented by Eq. (3b). In (5) the quantity $P_l$ $(\cos \theta_{\mu Z})$ is a spherical tensor of rank zero and so is expandable in terms of the inner product of spherical tensors of integral rank

$$P_l(\cos \theta_{\mu Z}) = \frac{4\pi}{2l+1} \sum_m (-1)^m Y_{l-m}(\theta_\mu, \varphi_\mu) Y_{lm}(\theta_p, \varphi_p).$$

Now $V_0$ will contain all terms with $l = 0$ so that the interaction Hamiltonian is

$$H' = \sum_{p=1}^{Z} \sum_{\substack{l \geq 1 \\ m}} (-1)^m f_l(r_p) g_l(r_\mu) Y_{l-m}(\theta_\mu, \varphi_\mu) Y_{lm}(\theta_p, \varphi_p) \tag{6}$$

where the radial functions are defined by

$$\left.\begin{aligned} f_l(r_p) &= e_p r_p^{\,l} \\ g_l(r_\mu) &= e_\mu / r_\mu^{l+1} \end{aligned}\right\} \quad r_p < r_\mu$$

$$\left.\begin{aligned} f_l(r_p) &= e_p / r_p^{l+1} \\ g_l(r_\mu) &= e_\mu r_\mu^{\,l} \end{aligned}\right\} \quad r_p > r_\mu. \tag{7}$$

(For a uniform charge distribution the spherically symmetric part of the potential has the form

$$V_0 = -\frac{Ze^2}{R_0}\left(\frac{3a_0^2}{2} - \frac{r}{2R_0^2}\right), \qquad r < R_0$$
$$= -\frac{Ze^2}{r}a_0^3, \qquad r > R_0$$

where $a_0$ and $R_0$ are defined in Eq. (2-2).)

Because of the interaction only the total angular momentum **F** and the parity are conserved where these are generally

$$\mathbf{F} = \mathbf{I} + \mathbf{j}_\mu$$
$$\pi = \pi_N \cdot \pi_\mu , \tag{8}$$

the nuclear and muon angular momenta and parity, respectively (for even-even nuclei $\mathbf{I} = \mathbf{L}$), and their associated magnetic quantum numbers are $\mathscr{M}$, $M_I$, and $m_\mu$. On using the usual relation between coupled and uncoupled representations (see Appendix A), the matrix elements of $H'$ are

$$\langle F'\mathscr{M}'I'j_\mu' \mid H' \mid F\mathscr{M}Ij_\mu\rangle = \delta_{FF'}\,\delta_{\mathscr{M}\mathscr{M}'} \sum_{p,l\geq 1} (-1)^{I'+j_\mu-F}$$
$$\times \sqrt{(2I'+1)(2I+1)}\; W(I, j_\mu, I', j_\mu'; Fl)$$
$$\times \langle I' \,||\, f_l(r_p)\, Y_l(\theta_p, \varphi_p) \,||\, I\rangle \langle j_\mu' \,||\, g_l(r_\mu)\, Y_l(\theta_\mu, \varphi_\mu) \,||\, j_\mu\rangle. \tag{9}$$

This is a generalization of expressions obtained earlier [111, 158]. The Racah coefficients $W(a, b, c, d; e, f)$ are defined in Rose [15].

In order to keep the discussion as general as possible we define a generalized penetration function $\mathscr{P}_l(I, I')$ in terms of the matrix elements of the electric multipole moment operators which are defined using the usual collective model assumptions of replacing $\Sigma_p\, e$ by $\int \varrho_0\, d\tau$ over the nuclear volume and so on,

$$\mathscr{E}^B_{lm} = \sqrt{\frac{16\pi}{2l+1}}\; \Sigma\; er'^l\, Y_{lm}(\theta_p', \varphi_p')$$
$$= \sqrt{\frac{16\pi}{2l+1}} \int \varrho_0(\mathbf{r}')\, r'^l\, Y_{lm}(\theta', \varphi')\, d\tau', \tag{10}$$

which is a generalization of Eq. (6-27) and (6-40). Then

$$\langle IM_I \mid \mathscr{E}^B_{lm} \mid I'M_I'\rangle \mathscr{P}_l(I, I') = C(I'lI;\ M_I',\ m,\ M_I) \sum_p \langle I \,||\, f_l\, Y_l{}^B \,||\, I'\rangle. \tag{11}$$

Now when $r_\mu > r_p$ then $f_l(r_p) = e_p r_p{}^l$ so that

$$\sum_p e_p f_l(r_p)\, Y^B_{lm} = \sqrt{\frac{2l+1}{16\pi}}\ \mathscr{E}^B_{lm}$$

hence

$$\mathscr{P}_l(I, I') = \sqrt{\frac{2l+1}{16\pi}}\,. \tag{12}$$

On the other hand, when $r_\mu < r_p$ then $f_l(r_p) = e_p / r_p^{l+1}$ and the penetration function will depend upon the nuclear model. For example, if one assumes a uniform charge distribution and a collective nuclear model of the type discussed in Chapter 3, then again Eq. (12) holds. It should be noted that these penetration functions are similar to those defined in Jacobsohn [111] except that we have placed their functional dependence on $r$ elsewhere.

Finally, we write the matrix elements of the interaction as

$$\begin{aligned}\langle F'\mathscr{M}'I'j_\mu' \mid H' \mid F\mathscr{M}Ij_\mu\rangle &= \delta_{FF'}\,\delta_{\mathscr{M}\mathscr{M}'} \sum_{l\geq 1} (-1)^{I'+j_\mu-F} \\ &\times \sqrt{(2I'+1)(2j_\mu'+1)}\ W(I,\ j_\mu,\ I',\ j_\mu';\ Fl)\,\langle I \,||\, D^l\, \mathscr{E}_l{}^B \,||\, I'\rangle \\ &\times \mathscr{P}_l(I',\ I)\,\langle j_\mu \,||\, g_l(r_\mu)\, Y_l(\theta_\mu,\ \varphi_\mu) \,||\, j_\mu'\rangle\,,\end{aligned} \tag{13}$$

in which we have used the Euler transformation of the spherical tensor $\mathscr{E}^L_{lm}$.

The terms in Eq. (13) associated with the electric quadrupole hyperfine interaction are those for which $l = 2$:

$$\begin{aligned}\langle F'\mathscr{M}'I'j_\mu' \mid H_2' \mid F\mathscr{M}Ij_\mu\rangle &= \delta_{FF'}\,\delta_{\mathscr{M}\mathscr{M}'}\,(-1)^{I'+j_\mu-F} \\ &\times \sqrt{(2I'+1)(2j_\mu'+1)}\ W(I,\ j_\mu,\ I',\ j_\mu';\ F2)\,\langle I \,||\, D^2\, \mathscr{E}_2{}^B \,||\, I'\rangle \\ &\times \mathscr{P}_2(I',\ I)\,\langle j_\mu \,||\, g_2(r_\mu) Y_2(\theta_\mu,\ \varphi_\mu) \,||\, j_\mu'\rangle.\end{aligned} \tag{14}$$

The reduced matrix elements involving the muon coordinates can be calculated as soon as the unperturbed problem of Eq. (3b) is solved. For a uniform charge distribution and a nuclear model of the type discussed in Chapter 3 the nuclear reduced matrix elements $\langle I \| D^2 \mathscr{E}_2{}^B \| I' \rangle$ are identical, to within a constant, to those given in Eq. (6-29). Thus they contain the effects of deformation vibrations and $K$-mixing. In his original investigation of this problem Wilets [158] used as a nuclear model that of Bohr and Mottelson so that only the terms in Eq. (6-29) diagonal in $K$ (associated with $Q_{00}$) appear in his interaction matrix elements and for even-even nuclei only $K = 0$ terms. In principle there is no reason to exclude from consideration off-diagonal matrix elements that involve the nuclear beta or "gamma" bands. For even-even nuclei Wilets' is quite a good assumption since all experiments to date have only been concerned with the splitting of the $K$ and $L$ mesic x rays.

The early work [111, 158] only considered the splitting of the $2P$ levels due to the interaction although in recent experiments it has been necessary to include all mesic levels for which $n \leq 4$ and the first two excited nuclear states ($I = 2$ and 4) [43, 85]. These experiments support the several details of the theory and furthermore, as can be seen from Eqs. (14) and (6-29), measure the sign of the quadrupole moment. In these nuclei the only other way to obtain this sign is from photonuclear experiments (see the previous chapter).

We turn now to consideration of odd-$A$ nuclei. Again the analogy with the electromagnetic transitions in such nuclei should prove useful. In Section D of Chapter 6 the electromagnetic operators contain two terms, one being associated with the core nucleons and the second with the extra-core particles. It is clear from Eqs. (5) and (6) that a similar situation pertains here so that for odd-$A$ nuclei the matrix elements of the interaction will contain a term associated with core-muon interactions and one for muon extra-core nucleon interactions. Confining our attention to the electric quadrupole interaction, we have

$$\begin{aligned}\langle F'\mathscr{M}'I'j_\mu' \mid H_2' \mid F\mathscr{M}Ij_\mu\rangle &= \delta_{FF'}\delta_{\mathscr{M}\mathscr{M}'}(-1)^{I'+j_\mu-F} \\ &\times \sqrt{(2I'+1)(2j_\mu'+1)}\, W(I, j_\mu, I', j_\mu'; F2)\, \langle j_\mu \| g_2(r_\mu) Y_2(\theta_\mu, \varphi_\mu) \| j_\mu' \rangle \\ &\times \{\langle I\chi \| D^2 f_2(r_0) Y_2(\theta, \varphi) \| I'\chi' \rangle + \langle I \| D^2 \mathscr{E}_2{}^B \| I' \rangle \mathscr{P}_2(I, I')\}. \qquad (15)\end{aligned}$$

Here, as in Eqs. (6-42), the first term in the brackets is the odd particle reduced matrix element of the interaction, the nuclear state functions being denoted by $| I\chi \rangle$ to make explicit the dependence on the Nilsson orbital. The second term in (15) is associated with the interaction with the collective degrees of freedom of the core and these interactions are identical to those discussed before for the even-even case. The only important difference is that the ground state spin $I$ is not zero and may be, in fact, quite large, giving rise to a very much more complicated hyperfine structure. Such effects have been measured in the odd-$A$ deformed nuclei [85] of $Ta^{181}$, $U^{235}$, and $Pu^{239}$. In the first of these it was found that only the static quadrupole hyperfine interaction was of importance which is consistent with the fact that while the first excited state is at 6.2 keV it is the head of the rotational band built on the 9/2 — [514] state. The ground state band is associated with the $\frac{7}{2}+$ [404] orbital with the associated 9/2 + level at 136.2 keV. In $U^{235}$ and $Pu^{239}$ dynamic interaction effects must be taken into account since the rotational structure is more compact with the first excited states in the ground state rotational band at 46 and 8 keV, respectively. (These ground state bands are formed from the $\frac{7}{2}-$ [743] and $\frac{1}{2}+$ [631] orbitals, respectively [48, 119].

The first term in Eq. (15) will, like its counterpart, discussed in Section D of Chapter 6, involve interactions within the ground state rotational band and also those which connect different single-particle orbitals. An example of this would be in the last-named nucleus $Pu^{239}$. The interaction would connect the ground state band orbital $\frac{1}{2}+$ [631] with the nearby $\frac{5}{2}+$ [622] orbital upon which is built a $K=\frac{5}{2}+$ rotational band [119]. No details of transitions involving such interactions have been worked out; however, it should be clear that to measure them we will have to examine the splittings of the associated mesic X-rays with great care and precision.

In this discussion we have not touched upon the very many other processes that occur when a muon is captured by a heavy atom, some of which have been investigated theoretically or experimentally. The most promising would seem to be the various nonradiative transitions, many of which will proceed through one collective level or another. We have already mentioned the mesic–Auger effect; however, this is known to be important only in fairly light atoms [148] and need not be discussed here.

One type of nuclear excitation that occurs in heavy nuclei and is certainly collective is that of fission. Wheeler originally pointed out that muons would induce fission by two distinctly different processes [41];

one involves the nuclear capture of the muon, which does not concern us here, and the other results from a nonradiative mesic transition. These are experimentally differentiable since the latter is essentially prompt, the former on a time scale of the order of the muon lifetime. Measurements of these processes indicate that about 5% of the muon-induced fissions in $U^{238}$ result from nonradiative mesic transitions [84]. It has been shown by Zaretski and Novikov [163] that this experiment can only be explained if the muon in the $K$ shell alters the size of the fission barrier. The more general problem of nonradiative $2P$–$1S$ muon transitions in the heavy elements has been investigated by these same authors [164]. For $Pu^{239}$ the effect is sizable since about 40% of the muons captured into an atomic system undergo a radiationless $2S$–$1S$ transition [45]. In tantalum the failure of these same experiments to detect any such transitions implies that the ratio of radiationless to radiative transitions in this atom is less than about 10%. Finally, we mention a suggestion that $E2$ radiationless transitions might well occur with a sufficient probability to be measured [144]. The process here is a radiationless $3D$–$1S$ transition that would compete with the radiative $3D$–$2P$ transition.

All of these ideas have only been investigated in the most general way wherein the details of nuclear structure and their experimental verification will probably have to await the arrival of so called "meson factories." However, it should seem clear that this nuclear probe might provide a most useful tool for the investigation of collective effects which, due to their relatively great excitation, are presently difficult to observe.

APPENDIX A

# Some Angular Momentum Theorems

It is useful to collect here some relations involving various angular momentum eigenfunctions; an adequate reference is Rose [15]. Throughout this volume the spherical harmonics $Y_{lm}(\theta, \varphi)$ have been assumed normalized in the sense

$$\int Y_{lm}(\theta, \varphi) Y_{l'm'}(\theta, \varphi)\, d\omega = \delta_{ll'}\, \delta_{mm'},$$

$d\omega = \sin\theta\, d\theta\, d\varphi$ with the integration being over the unit sphere.

The transformation between laboratory and body-fixed coordinate systems is effected by the rotation matrices, $D^I_{MK}$ which are the $(2I+1)$-dimensional representation of the rotation group $R(3)$

$$| I, K\rangle^B = \sum_M D^I_{MK}(\theta_i)\, | I, M\rangle^L,$$

where the $\theta_i$ are the Euler angles specifying the transformation.

By labeling the laboratory coordinate system $(x, y, z)$ and the body-fixed coordinate system (1,2, 3), then the three Euler angles used here are

(i) A rotation through $\theta_1$ about the $z$ axis yielding the first intermediate coordinate system $x'$, $y'$, $z' = z$.

(ii) A rotation through $\theta_2$ about the $y'$ axis yielding the second intermediate coordinate system $x''$, $y''$, $= y'$, $z''$.

(iii) A rotation through $\theta_3$ about the $z''$ axis yielding the body-fixed system. All rotations are considered positive if the usual right-hand screw when so rotated advances along the axis of rotation in the positive direction. The full transformation matrix is to be found in Whittaker [19], although his angle labels are different from these.

The functions $D^{I*}_{M,K}(\theta_i)$ are simultaneous eigenfunctions of the square of the total angular momentum, $I^2$, of the projection of the angular momentum on the $z$ axis, $I_z$, and on the 3 axis, $I_3$, with eigenvalues $I(I+1)$, $M$ and $K$, respectively. Thus they are the appropriate eigenfunctions for the symmetric top whose symmetry axis is coincident with the 3 axis. Since the $D^I_{MK}(\theta_i)$ are the elements of a unitary transformation they possess the usual orthogonality properties

$$\sum_{M=-I}^{I} D^{I*}_{MK}(\theta_i)\, D^I_{MK'}(\theta_i) = \delta_{KK'}$$

$$\sum_{K=-I}^{I} D^{I*}_{MK}(\theta_i)\, D^I_{M'K}(\theta_i) = \delta_{MM'}.$$

Also,

$$D^I_{MK}(-\theta_3, -\theta_2, -\theta_1) = D^{I*}_{KM}(\theta_i),$$

since the set of rotations $(-\theta_3, -\theta_2, -\theta_1)$ form the inverse transformation to the set $(\theta_1, \theta_2, \theta_3)$. Also

$$D^{I*}_{MK}(\theta_i) = (-1)^{M-K}\, D^I_{-M-K}(\theta_i).$$

A useful analytic form for these functions due to Wigner [21] is

$$D^I_{MK}(\theta_1, \theta_2, \theta_3) = e^{-iM\theta_1} e^{-iK\theta_3} [(I+K)!(I-K)!(I+M)!(I-M)!]^{1/2}$$
$$\times \sum_j (-1)^j [(I-M-j)!\,(I+K-j)!(j+M-K)!\,j!]^{-1}$$
$$\times (-\sin \tfrac{1}{2}\theta_2)^{M-K+2j} (\cos \tfrac{1}{2}\theta_2)^{2(I-j)+K-M}.$$

To determine the action the relabeling transformations $T_1$, $T_2$ of Chapter 3 have upon these functions note that

(i) $$T_1(\varphi_1, \varphi_2, \varphi_3)\, D^I_{MK}(\theta_i) = \sum_{K'} D^I_{MK'}(\theta_i)\, D^I_{K'K}(\theta_i).$$

Since the rotation is $\pi$ about the 2 axis

$$\varphi_1 = \varphi_3 = 0$$

$$\varphi_2 = \pi$$

and using the analytic form of the $D^I_{MK}$ functions

$$D^I_{MK}(0, \pi, 0) = (-1)^{I+K} \delta_{M-K}$$

thus

$$T_1 D^{I*}_{MK}(\theta_i) = (-1)^{I-K} D^{I*}_{M-K}(\theta_i).$$

(ii) Since $T_2^2$ is two successive rotations of $\pi/2$ about the 3 axis clearly

$$D^I_{MK}(00\pi) = e^{-iK\pi} \delta_{KM}$$

whence

$$T_2^2 D^{I*}_{MK} = e^{iK\pi} D^{I*}_{MK} .$$

Let the total angular momentum $\mathbf{I}$ be the sum of two independent angular momenta $\mathbf{I}_1$ and $\mathbf{I}_2$, that is

$$\mathbf{I} = \mathbf{I}_1 + \mathbf{I}_2$$

where

$$[\mathbf{I}_1 , \mathbf{I}_2] = 0.$$

Consider two angular momentum representations: the uncoupled representation $| I_1 , I_2 , m_1 , m_2 \rangle$ where

$$I_i^2 \mid I_1 , I_2 , m_1 , m_2 \rangle = I_i(I_i + 1) \mid I_1 , I_2 , m_1 , m_2 \rangle$$

$$I_{iz} \mid I_1 , I_2 , m_1 , m_2 \rangle = m_i \mid I_1 , I_2 , m_1 , m_2 \rangle$$

and the coupled representation $| I_1 , I_2 , I , m \rangle$ where

$$I^2 \mid I_1 , I_2 , m \rangle \quad = I(I + 1) \mid I_1 , I_2 , I, m \rangle$$

$$I_z \mid I_1 , I_2 , I, m \rangle = m \mid I_1 , I_2 , I, m \rangle.$$

These two representations are related by

$$| I_1 , I_2 , I, m \rangle = \sum_{m_1} C(I_1 , I_2 , I; m_1 , m - m_1 , m) \mid I_1 , I_2 , m_1 , m - m_1 \rangle$$

the phases of the Clebsch–Gordan coefficients being fixed by the requirement

$$C(I_1 , I_2 , I_1 + I_2; I_1 , I_2 , I_1 + I_2) = 1.$$

Also $C(I_1 , I_2 , I; m_1 , m_2 , m) = 0$ unless $m = m_1 + m_2$ and $| I_1 - I_2 |$

$\leq I \leq I_1 + I_2$. A number of symmetry relations are useful

$$C(I_1, I_2, I; -m_1, -m_2, -m) = (-1)^{I_1+I_2-I}\, C(I_1, I_2, I; m_1, m_2, m)$$
$$C(I_2, I_1, I; m_2, m_1, m) = (-1)^{I_1+I_2-I}\, C(I_1, I_2, I; m_1, m_2, m)$$
$$C(I_1, I, I_2; m_1, -m, -m_2) = (-1)^{I_1-m_1} \sqrt{\frac{2I_2+1}{2I+1}}\, C(I_1, I_2, I; m_1, m_2, m).$$

Others may be derived from these. The orthogonality conditions are:

$$\sum_{m_1} C(I_1, I_2, I; m_1, m-m_1, m)\, C(I_1, I_2, I'; m_1, m-m_1, m) = \delta_{II'}$$
$$\sum_{I} C(I_1, I_2, I; m_1, m-m_1, m)\, C(I_1, I_2, I; m_1', m'-m_1', m') = \delta_{m_1 m_1'}\, \delta_{mm'}.$$

An important relation between the product of two $D^I_{MK}(\theta_i)$ functions and a single one is

$$D^{I_1}_{M_1K_1}(\theta_i)\, D^{I_2}_{M_2K_2}(\theta_i) = \sum_I C(I_1, I_2, I; M_1, M_2, M_1+M_2) \times C(I_1, I_2, I; K_1, K_2, K_1+K_2)\, D^{I(\theta_i)}_{M_1+M_2,\, K_1+K_2}(\theta_i).$$

This is known as the Clebsch–Gordan series. The inverse relation can be obtained by using the orthogonality relations of the Clebsch–Gordan coefficients.

The orthogonality and normalization of the $D^I_{MK}(\theta_i)$ are obtained from

$$\int D^{I_1*}_{M_1K_1}(\theta_i)\, D^{I_2}_{M_2K_2}(\theta_i)\, d\Omega = \frac{8\pi^2}{2I_1+1}\, \delta_{I_1,I_2}\, \delta_{M_1,M_2}\, \delta_{K_1,K_2}.$$

Here $d\Omega = d\theta_1 \sin\theta_2\, d\theta_2\, d\theta_3$ with $0 \leq \theta_1 \leq 2\pi$, $0 \leq \theta_2 \leq \pi$, $0 \leq \theta_3 \leq 2\pi$. To calculate transition probabilities involving top state functions one finds integrals involving three of the $D^I_{MK}(\theta_i)$

$$\int D^{I_1*}_{M_1K_1}(\theta_i)\, D^{I_2}_{M_2K_2}(\theta_i)\, D^{I_3}_{M_3K_3}(\theta_i)\, d\Omega$$
$$= \frac{8\pi^2}{2I_1+1} C(I_3, I_2, I_1; M_3, M_2, M_3+M_2)\, C(I_3, I_2, I_1; K_3, K_2, K_3+K_2) \times \delta_{M_1, M_2+M_3}\, \delta_{K_1, K_2+K_3}.$$

The relation between the spherical harmonics $Y_{l,m}(\theta, \varphi)$ and the $D^I_{MK}(\theta_i)$ is

$$Y_{lm}(\theta, \varphi) = \sqrt{\frac{2l+1}{4\pi}}\, D^{l*}_{M0}(\theta, \varphi, 0).$$

The time derivatives of the $D^L_{MK}(\theta_i)$ are most easily gotten by noting that these functions are just the matrix elements of the rotation operator $e^{-i\boldsymbol{\theta}\cdot\mathbf{L}}$ where the vector $\boldsymbol{\theta}$ is in the direction of that rotation bringing laboratory and body-fixed axes into coincidence, $|\,\boldsymbol{\theta}\,|$ being its magnitude. On calling $\theta_k$ the projections of $\boldsymbol{\theta}$ on the body-fixed axes and $\omega_k$ the similar angular velocities

$$\dot{D}^{L*}_{MK} = \frac{d}{dt}\langle LK \mid e^{i\boldsymbol{\theta}\cdot\mathbf{L}} \mid LM\rangle = i\sum_{k=1}^{3} \omega_k \langle LM \mid L_k\, e^{i\theta\cdot\mathbf{L}} \mid LM\rangle$$

from which the expression (14) of Chapter 2 follows at once.

Finally, we wish to determine the angular momentum commutation relations in the body-fixed system. We do this in a straightforward manner. The angular velocity of the body, referred to the body-fixed system [19] is

$$\begin{aligned}
\omega_1 &= \dot{\theta}_2 \sin\theta_3 + \dot{\theta}_1 \sin\theta_2 \cos\theta_3 \\
\omega_2 &= \dot{\theta}_2 \cos\theta_3 - \dot{\theta}_1 \sin\theta_2 \sin\theta_3 \\
\omega_3 &= \dot{\theta}_3 - \dot{\theta}_1 \cos\theta_2\,,
\end{aligned}$$

where as usual $\dot{\theta}_i = \partial\theta_i/\partial t$ and the Euler angles $\theta_i$ have been defined before. If $\mathscr{I}$ be the inertial tensor then the angular momentum of the body is defined as

$$\mathbf{L} = \mathscr{I}\,\boldsymbol{\omega}\,,$$

which we must express in terms of the coordinates and conjugate momenta.

For a rigid body moving with one point fixed the Lagrangian $\mathscr{L}$ and kinetic energy $T$ functions are identical

$$\mathscr{L} = (1/2)\,\boldsymbol{\omega}\,\mathscr{I}\,\boldsymbol{\omega}$$

and the conjugate momenta follow from the definition

$$P_i = \partial\mathscr{L}/\partial\dot{q}_i.$$

From this we find upon solving for the angular momenta

$$L_1 = P_{\theta_2} \sin \theta_3 + (P_{\theta_1} -- P_{\theta_3} \cos \theta_2) \cos \theta_3/\sin \theta_2$$
$$L_2 = P_{\theta_2} \cos \theta_3 - (P_{\theta_1} - P_{\theta_3} \cos \theta_2) \sin \theta_3/\sin \theta_2$$
$$L_3 = P_{\theta_3}.$$

To calculate the commutator $[L_1, L_2]$ we must evaluate commutators of the form $[P_2, \cos \theta_2]$ which is most easily done making use of the quantum condition relating the classical Poisson bracket with the commutator [7]

$$[A, B] = i\hbar\{A, B\} = i\hbar \sum_{i=1}^{n} \left( \frac{\partial A}{\partial q_1} \frac{\partial B}{\partial P_i} - \frac{\partial B}{\partial q_i} \frac{\partial A}{\partial P_i} \right).$$

Thus we get

$$[L_1, L_2] = -i\hbar L_3, \text{ cyclically.}$$

From this relation it is clear that the operator

$$L_+{}^B = L_1 + iL_2$$

steps the value of $K$ *down* by one unit whereas the conjugate operator

$$L_-{}^B = L_1 - iL_2$$

steps the value of $K$ *up* by one unit when applied to an angular momentum state function.

We may define a body-fixed, pseudo-spherical coordinate system such that the components of the angular momentum operator are

$$L_{\pm 1} = \mp (1/\sqrt{2})L_{\pm}{}^B$$
$$L_0 = L_3.$$

Thus in the laboratory coordinate system

$$[L_M, L_N] = -\sqrt{2}\; C(111, M, N, M + N)L_{M+N},$$

whereas in the body-fixed coordinate system

$$[L_M, L_N] = +\sqrt{2}\; C(111; M, N, M + N)L_{M+N}.$$

APPENDIX B

# Collective Parameters

In this appendix we shall outline the derivations of certain collective parameters. First, we shall discuss the condition to fix the center of mass, then the volume conserving terms of the nuclear radius, and finally derive, for oscillations about spherical, the potential energy parameter $C_\lambda$. The ranges of the asymmetry parameters are also obtained.

We expand the nuclear radius in spherical harmonics of all orders in the body-fixed coordinate system as

$$R^B(\theta', \varphi') = R_0[a_0{}^* + \sum_{\lambda,\mu} a^*_{\lambda\mu} Y_{\lambda\mu}(\theta', \varphi')]. \tag{1}$$

The center of mass is defined as

$$M\mathbf{R} = \sum_i m_i \mathbf{r}_i$$

the sum being over the number of particles. For purposes of illustration we take the $z$ component and for a collective calculation make the replacement

$$\sum_i m_i \rightarrow \varrho_m \, dv$$

where the mass density $\varrho_m$ is just

$$\varrho_m = \frac{3M}{4\pi R_0{}^3}$$

assuming a uniform mass distribution. Since

$$Z = r\cos\theta = \sqrt{\frac{4\pi}{3}}\, r\, Y_{10}(\theta', \varphi')$$

we find

$$Z = \sqrt{\frac{3}{4\pi}}\,\frac{1}{R_0^3}\int Y_{10}^*(\theta', \varphi')\, r^3\, dr\, d\Omega$$

$$= \sqrt{\frac{3}{4\pi}}\,\frac{R_0}{4}\int Y_{10}^*(\theta', \varphi')[a_0^* + \sum_{\lambda\mu} a_{\lambda\mu}^* Y_{\lambda\mu}(\theta', \varphi')]^4\, d\Omega$$

$$= \sqrt{\frac{3}{4\pi}}\, R_0 a_0^{*3}\, a_{10}^* \tag{2}$$

to first order. Now $a_0$ is of the order unity so that if (2) is to vanish $a_{10}^*$ must. In a similar manner the $x$ and $y$ components yield this same condition on the parameters $a_{1\pm1}^*$.

The constant volume condition is just

$$\int r^2\, dr\, d\Omega = \frac{4}{3}\pi R_0^3$$

whence we find in the laboratory system

$$\frac{1}{4\pi}\int [\alpha_0^* + \sum_{\substack{\lambda>1\\ \mu}} \alpha_{\lambda\mu}^* Y_{\lambda\mu}(\theta, \varphi)]^3\, d\Omega$$

$$= \alpha_0^{*3} + \frac{3\alpha_0^*}{4\pi}\sum_{\substack{\lambda>1\\ \mu}} |\alpha_{\lambda\mu}|^2 + \frac{1}{4\pi}\mathscr{E} + \cdots = 1 \tag{3}$$

where the third-order term is

$$\mathscr{E} = \sum_{\lambda,\lambda',\lambda''}\sqrt{\frac{(2\lambda''+1)(2\lambda'+1)}{4\pi(2\lambda+1)}}\, C(\lambda''\lambda'\lambda;\, 0,\, 0,\, 0)$$
$$\times \sum_{\mu,\mu',\mu''} C(\lambda''\lambda'\lambda;\, \mu''\mu'\mu)\alpha_{\lambda\mu}^*\alpha_{\lambda'\mu'}\alpha_{\lambda''\mu''} \tag{4}$$

which vanishes unless $\lambda'' + \lambda' + \lambda$ is even. To proceed we substitute into (3)

$$\alpha_0^* = 1 - \delta$$

which through third order

$$\delta = \frac{1}{4\pi} \sum_{\substack{\lambda>1 \\ \mu}} | \alpha_{\lambda\mu} |^2 + \frac{1}{12\pi} \mathscr{C}$$

so that to first order $\alpha_0{}^*$ is unity. The nuclear radius is then

$$R^L(\theta, \varphi) = R_0 \left[ 1 - \frac{1}{4\pi} \sum_{\substack{\lambda>1 \\ \mu}} | \alpha_{\lambda\mu} |^2 - \frac{1}{12\pi} \mathscr{C} + \sum_{\substack{\lambda>1 \\ \mu}} \alpha^*_{\lambda\mu} Y_{\lambda\mu}(\theta, \varphi) \right]. \quad (5)$$

We now turn to the potential energy which is the sum of coulomb and surface energies

$$V = V_c + V_s ,$$

and calculate the coulomb energy first. We calculate the electrostatic potential at a point $(r, \theta, \varphi)$ exterior to the charged drop. If $\varrho$ be the charge of a unit volume $dv'$ located at $(r', \theta', \varphi')$ and a distance $\mathbf{R} = \mathbf{r} - \mathbf{r}'$ from the field point then the potential is

$$V_c(r) = \varrho_c \int_\tau \frac{dv'}{R}$$

$$= 4\varrho_c \sum_{l,m} \frac{1}{(2l + 1)r^{l+1}} Y_{l,m}(\theta, \varphi) \int_\tau r'^l \, Y^*_{lm}(\theta', \varphi') \, dv' .$$

Here $\tau$ indicates integration over the nuclear volume. Carrying out the integration and substituting $\varrho_c = 3Ze/4\pi R_0{}^3$, we find the electrostatic potential in the region outside the drop to be

$$V_c(r) = Ze \left[ \frac{1}{r} + \sum_{\lambda\mu} \frac{3R_0{}^\lambda}{(2\lambda + 1)r^{\lambda+1}} \alpha^*_{\lambda\mu} Y_{\lambda\mu}(\theta, \varphi) \right],$$

which is to be evaluated at the nuclear surface and is

$$V_c(R_0) = \frac{Ze}{R_0} \left[ 1 - 2 \sum_{\lambda\mu} \frac{\lambda - 1}{2\lambda + 1} \alpha^*_{\lambda\mu} Y_{\lambda\mu}(\theta, \varphi) \right] \quad (6)$$

to first order.

We now wish to calculate the difference between the work necessary to assemble the deformed system and the spherical one. We may consider the former to be built up like an onion, each layer having some thickness $\Delta R$ but the same angular shape. That is the free surface at any time is in terms of an assembly parameter $k$

$$R(\theta,\ \varphi,\ k) = R_0[1 + k \sum_{\lambda\mu} \alpha^*_{\lambda\mu} Y_{\lambda\mu}(\theta,\ \varphi)],$$

so that the range of $k$ is $0 \leq k \leq 1$. The thickness of an onion layer is

$$dR = R_0 \sum_{\lambda\mu} \alpha^*_{\lambda\mu} Y_{\lambda\mu}(\theta,\ \varphi)\, dk.$$

Now $d\Omega$ cuts out of the drop an element of volume $R^2 dR\, d\Omega$ with charge $dq$

$$\begin{aligned} dq &= \varrho_c R^2\, dR\, d\Omega \\ &= \varrho_c R_0{}^3 \sum_{\lambda\mu} \alpha^*_{\lambda\mu} Y_{\lambda\mu}(\theta,\ \varphi)\, d\Omega\, dk. \end{aligned} \tag{7}$$

To obtain the work necessary to assemble the deformed system one integrates the increment of work $dW$ necessary to add a layer of charge which is the potential (6) times $dq$. In Eq. (6) the constant term, $Ze/R_0$, is the potential of a sphere of radius $R_0$ so that $k$ times the remaining term is the potential difference between the sphere and deformed shape. That is

$$d(V_c - V_c{}^0) = -\frac{2Ze}{R_0}\, k \sum_{\lambda\mu} \frac{\lambda - 1}{2\lambda + 1} \alpha^*_{\lambda\mu} Y_{\lambda\mu}(\theta,\ \varphi)\, dq$$

whence

$$V_c = V_c{}^0 - \frac{3Z^2e^2}{4\pi R_0} \sum_{\lambda\mu} \frac{\lambda - 1}{2\lambda + 1} \mid \alpha_{\lambda\mu} \mid^2. \tag{8}$$

Now the $\alpha_{\lambda\mu}$ are the generalized coordinates of the problem and in this case they are also the normal coordinates. If $\xi_i$ are the normal coordinates of a problem it is well known that the potential function can be written as

$$\begin{aligned} U &= U_0 + \frac{1}{2} \sum_{ij} b_{ij} \xi_i \xi_j \\ &= U_0 + \frac{1}{2} \sum_{i} b_i \xi_i{}^2. \end{aligned} \tag{9}$$

From Eqs. (8) and (9) we may identify the coulomb potential energy parameter

$$b_\lambda^c = -\frac{3Z^2e^2}{2\pi R_0} \cdot \frac{(\lambda - 1)}{(2\lambda + 1)} . \tag{10}$$

Consider now the energy due to the distortion of the surface. Starting with a spherical surface we carry out an arbitrary infinitesimal displacement $\delta\mathbf{r}$

$$\begin{aligned} \delta\mathbf{r} &= \delta\hat{\mathbf{r}} + \delta\hat{\boldsymbol{\theta}} + \delta\hat{\boldsymbol{\varphi}} \\ &= (\hat{\mathbf{r}} + \hat{\boldsymbol{\theta}} + \hat{\boldsymbol{\varphi}})\, \delta\varkappa \end{aligned}$$

where $\hat{\mathbf{r}}$, $\hat{\boldsymbol{\theta}}$, $\hat{\boldsymbol{\varphi}}$ are the unit vectors and $\delta\varkappa$ an infinitesimal scalar. The variation in the surface energy is the scalar product of the surface tension by $\delta\mathbf{r}$. For a small element of surface

$$d\mathbf{F}_S = -\mathscr{S}\, d\mathbf{S} = -\mathscr{S}\, dS\, \hat{\mathbf{n}}_S \tag{11}$$

$\hat{\mathbf{n}}_s$ being the unit vector normal to the surface at $dS$ and $\mathscr{S}$ the coefficient of surface tension. The expression (11) is negative as the surface tension is directed inward. From the laboratory expansion of the nuclear surface

$$R^L = R_0[1 + \sum_{\lambda\mu} \alpha^*_{\lambda\mu} Y_{\lambda\mu}(\theta, \varphi)] \tag{12}$$

we obtain

$$dR - \frac{1}{R} \frac{\partial R}{\partial \theta} R\, d\theta - \frac{1}{R \sin\theta} \frac{\partial R}{\partial \varphi} R \sin\theta\, d\varphi = 0.$$

So that the surface has direction cosines

$$\frac{1}{\mathscr{R}}, \quad \frac{-\dfrac{1}{R}\dfrac{\partial R}{\partial \theta}}{\mathscr{R}}, \quad \frac{-\dfrac{1}{R\sin\theta}\dfrac{\partial R}{\partial \varphi}}{\mathscr{R}},$$

where

$$\mathscr{R}^2 = 1 + \frac{1}{R^2}\left(\frac{\partial R}{\partial \theta}\right)^2 + \frac{1}{R^2 \sin\theta^2}\left(\frac{\partial R}{\partial \varphi}\right)^2$$

Thus

$$\begin{aligned} \delta\, dE(s, \varkappa) &= dE(s)\, \delta\varkappa = -\mathscr{S}\, d\mathbf{s} \cdot \delta\mathbf{r} \\ &= -\frac{\mathscr{S}}{\mathscr{R}} R^2\left(1 - \frac{1}{R}\frac{\partial R}{\partial \theta} - \frac{1}{R\sin\theta}\frac{\partial R}{\partial \varphi}\right) d\Omega\, d\varkappa. \end{aligned}$$

Now $dE(S)$ is given immediately since $\varkappa$ is arbitrary. To proceed expand $\mathscr{R}$ keeping only the lowest order terms, whence

$$E(s) = \int_s dE(s) = -\mathscr{S} \int_s \left\{R^2 - \frac{1}{2}\left[\left(\frac{\partial R}{\partial \theta}\right)^2 + \frac{1}{\sin^2\theta}\left(\frac{\partial R}{\partial \varphi}\right)^2\right]\right\} d\Omega$$

the integration being over the nuclear surface. If the surface energy of the sphere is $E_S{}^0$ then we seek

$$E(s) - E_s{}^0 = -\mathscr{S} \int_s \left\{R^2 - R_0{}^2 - \frac{1}{2}\left[\left(\frac{\partial R}{\partial \theta}\right)^2 + \frac{1}{\sin^2\theta}\left(\frac{\partial R}{\partial \varphi}\right)^2\right]\right\} d\Omega.$$

Substituting Eq. (12) into this and integrating yields

$$E(s) - E_s{}^0 = \frac{\mathscr{S} R_0{}^2}{2} \sum_{\lambda\mu} (\lambda - 1)(\lambda + 2) \mid \alpha_{\lambda\mu} \mid^2. \tag{13}$$

Again the comments about normal coordinates made before apply here so that we identify the surface potential energy parameter

$$b_\lambda{}^S = (\lambda - 1)\,(\lambda + 2)\mathscr{S} R_0{}^2. \tag{14}$$

The sum of Eqs. (10) and (14) is the potential energy parameter $C_\lambda$ of Eq. (2-9).

To augment the discussion in Chapter 3 on the symmetry requirements of the various models imposed by invariance under the general relabeling transformations $T_i$, we note that for the deformed quadrupole surfaces the excursion of the surface along the several principal axes is given by

$$\delta R_\varkappa = \sqrt{\frac{5}{4\pi}}\, \beta\, R_0 \cos\left(\gamma - \frac{2\pi}{3}\varkappa\right). \tag{15}$$

Clearly $\beta$ is unchanged by $T_1$, $T_2$, and $T_3$; however, from Eq. (15) it is seen that

$$T_1\gamma = \gamma$$

$$T_2\gamma = -\gamma$$

$$T_3\gamma = \left(\gamma - \frac{2\pi}{3}\right).$$

From these it follows that

$$H(-\gamma) \quad = H(\gamma) \tag{16a}$$

$$H\left(\gamma - \frac{2\pi}{3}\right) = H(\gamma) \tag{16b}$$

$$H\left(\frac{2\pi}{3} - \gamma\right) = H(\gamma). \tag{16c}$$

Equation (16c) is implied by Eqs. (16a, b). From these we see that the effective range of $\gamma$ is $0 \leq \gamma \leq 2\pi/3$ and the solutions are symmetric about $\gamma = \pi/3$. Indeed, $\gamma$ is even more restricted because of the form of the moments of inertia and the fact that

$$\sin^2\left(\gamma - \frac{2\pi}{3}\varkappa + \frac{\pi}{3}\right) = \sin^2\left(\gamma - \frac{2\pi}{3}\varkappa'\right), \qquad \varkappa, \varkappa' = 1, 2, 3,$$

the more restrictive relations

$$H\left(\gamma \pm \frac{\pi}{3}\right) = H(\gamma) \tag{17a}$$

$$H\left(\frac{\pi}{3} - \gamma\right) = H(\gamma), \tag{17b}$$

apply instead of (16b) and (16c), respectively. It is for this reason that $\gamma$ is limited to the range $0 \leq \gamma \leq \pi/6$.

Similar considerations apply to the negative parity models. Here it is possible to use the form of the moments of inertia directly. We consider only the case discussed in Chapter 3 where the octupole asymmetry parameter $\iota = 0$. Then from the form of the moments of inertia of Eq. (3-56) we observe that

$$H(\eta \pm \pi) = H(\eta) \tag{18a}$$

$$H(\pi - \eta) = H(\eta) \tag{18b}$$

from which it follows that the effective range of $\eta$ is $0 \leq \eta \leq \pi$ and further the solutions of the octupole rotor problem are symmetric about $\pi/2$.

# APPENDIX C

# Model State Functions

As a matter of convenience we list here the state functions for the various models discussed.

## A. The Vibrational Model

The vacuum state function $| 0\rangle$ represents the ground state of an even-even nucleus. Surfon creation and destruction operators $b^+_{\lambda\mu}$ and $b_{\lambda\mu}$, respectively, are defined by

$$[b_{\lambda\mu}, b^+_{\lambda'\mu'}] = \delta_{\lambda\lambda'} \delta_{\mu\mu'},$$

all others are zero and

$$b_{\lambda\mu} | 0\rangle = 0$$
$$b^+_{\lambda\mu} | 0\rangle = | \lambda 1 \lambda\mu\rangle$$

with the state functions denoted by $| \lambda N_\lambda L M_L\rangle$. The first four states are then

$$| 0\rangle = | \lambda 000\rangle$$
$$| \lambda 1 \lambda\mu\rangle = b^+_{\lambda\mu} | 0\rangle$$
$$| \lambda 2 L\mu\rangle = \frac{1}{\sqrt{2}} \sum_{\nu} C(\lambda\lambda L;\ \nu,\ \mu - \nu,\ \mu)\, b^+_{\lambda\mu-\nu} b^+_{\lambda\nu} | 0\rangle$$

$$| \lambda 3L\mu\rangle = \left[2 + 4(2L' + 1) \begin{Bmatrix} \lambda\lambda L \\ L'\lambda L \end{Bmatrix}\right]^{-1/2}$$

$$\times \sum_{\nu\sigma} C(\lambda LL'; \sigma, \mu - \sigma, \mu) C(\lambda\lambda L; \nu, \sigma - \nu, \sigma) b^+_{\lambda\mu-\sigma} b^+_{\lambda\mu-\nu} b^+_{\lambda\nu} | 0\rangle$$

where { } is a six-$j$ symbol [6] and the intermediate angular momentum $L'$ must be specified. As is usual in coupling up these state functions, states differing only in the intermediate angular momentum are not orthogonal. Higher state functions may be constructed in a similar fashion; however, a good deal of care must be exercised in determining the normalization constants.

## B. Deformed Even–Even Nuclei with Axial Symmetry

Axial symmetry implies that $K$ is a good quantum number; thus for positive parity states the total state function can be written

$$| n_\beta n_\gamma LMK\rangle = \Phi_{n_\beta n_\gamma}(\beta, \gamma) | LMK\rangle$$

where

$$| LMK\rangle = \sqrt{\frac{2L + 1}{16\pi^2}} (D^{L*}_{MK}(\theta_i) + (-1)^L D^{L*}_{M-K}(\theta_i)).$$

For small oscillations about the axially symmetric shape, $\Phi_{n_\beta n_\gamma}(\beta, \gamma)$ separates into

$$\Phi_{n_\beta n_\gamma}(\beta, \gamma) = f(\beta) g(\gamma)$$

where

$$f(\beta) = N_\beta H_{n_\beta}(\sqrt{B_2\omega_\beta/\hbar} [\beta - \beta_0]) \exp(-B_2\omega_\beta[\beta - \beta_0]^2/2\hbar)$$
$$g(\gamma) = N_\gamma \gamma^{-1} W_{(n_\gamma+1)/2, |K|/4}(\sqrt{B_\gamma C_\gamma}\, \gamma^2/\hbar)$$

where $H_n$ is a Hermite polynomial of order $n$ and $W_{k,m}(z)$ is the Whittaker function. The normalization is determined by

$$\int_{-\infty}^{\infty} f^2(\beta)\, d\beta = 1$$

$$\int g^2(\gamma) | \gamma | d\gamma = 1,$$

respectively.

Upon coupling the octupole negative parity oscillations, the total vibrational wave function has the form

$$\Psi_{n_\beta, n_b, n_\gamma, n_g}(\beta, b, \gamma, g) = f(\beta)F(b)G(\gamma, g),$$

where $f(\beta)$ is the function defined before for the pure quadrupole case and

$$F(b) = N_b H_{n_b}(\sqrt{B_3\omega_b/\hbar}\, b) \exp(-B_3\omega_b b^2/2\hbar).$$

Finally, the function $G(\gamma, g)$ can be approximately separated by introducing the variables $\Gamma$ and $\sigma$

$$\gamma = \Gamma \cos \sigma$$
$$g = \Gamma \sin \sigma.$$

Then to lowest order

$$G(\gamma, g) = P(\Gamma)S(\sigma).$$

The function $S(\sigma)$ satisfies Eq. (3-52b) with solutions

$$S(\sigma) = N_\sigma e^{i\mu\sigma}, \qquad \mu \text{ integral}$$

while $P(\Gamma)$ satisfies Eq. (3-52a) with solutions

$$P(\Gamma) = N_\Gamma\, \Gamma^\lambda \exp(-\tfrac{1}{2} C_\gamma \Gamma^2)\, L^{\lambda+1/2}_{1/2(n_\Gamma-\lambda)}(C_\gamma \Gamma^2).$$

Here $L_n{}^m(x)$ is an associated Laguerre polynomial and the other quantities have been defined in Chapter 3, Section D.

## C. Deformed Even–Even Nuclei without Axial Symmetry.

Since the solution of this problem with all $(2\lambda + 1)$ degrees of freedom is identical with that given in Section A we consider only systems with four degrees of freedom, three rotational, and one vibrational,

$$| n_{\beta_\lambda}, LM\pi\rangle = \Phi^{(\lambda)}_{LNn}(\beta_\lambda)\, | LM\pi\rangle$$

where the rotational part contains a sum over the projection quantum number $K$ and we consider only models of positive and negative parity with the lowest $\lambda$.

1. *Positive parity* $\lambda = 2$.

$$| LM +\rangle = \sqrt{\frac{2L+1}{16\pi^2}} \sum_{K=0}^{L} A_K^{L+}(\gamma) \, [D_{MK}^{L*}(\theta_i) + (-1)^L D_{M-K}^{L*}(\theta_i)],$$

which belongs to the representation $A$.

2. *Negative parity* $\lambda = 3$.

$$| LM -\rangle = \sqrt{\frac{2L+1}{16\pi^2}} \sum_{K=0}^{L} A_K^{L-}(\eta) \, [D_{MK}^{L*}(\theta_i) - (-1)^L D_{M-K}^{L*}(\theta_i)],$$

which belongs to the representation $B_1$.

The vibrational solutions are the same for all $\lambda$

$$\Phi_{LNn}^{(\lambda)}(\beta_\lambda) = \beta_\lambda^{-3/2} D_\nu(\beta_\lambda),$$

where the $D_\nu(\beta_\lambda)$ are Weber's parabolic cylinder functions being solutions to

$$\left[\frac{d^2}{dy^2} + 2\nu + 1 - y^2\right] D_\nu \, (\sqrt{2}\, y) = 0,$$

subject to the boundary conditions

$$\lim_{y \to \infty} D_\nu(\sqrt{2}\, y) = 0$$

$$D_\nu(-\sqrt{2}\, Z_1) = 0$$

with

$$y = Z_1 \frac{\beta_\lambda - \beta_\lambda(L, N)}{\beta(L, N)}$$

$$Z_1{}^4 = \left[\frac{\beta_\lambda(L, N)}{\beta_{\lambda 0}\mu}\right]^4 + \frac{3}{2}\left(\mathscr{E}_{L,N}^{(\lambda)} + \frac{3}{2}\right).$$

## D. Deformed Odd–$A$ Nuclei

For the systems with axial symmetry and in the strong coupling approximation the particle and rotational solutions essentially separate and $K$ and $\Omega$, the rotor and particle symmetry axis angular momentum projections, are good quantum numbers. Furthermore, the solutions are restricted to the case $K = \Omega$. The properly symmetrized state functions are

$$|EIMK\rangle = \sqrt{\frac{2I+1}{16\pi^2}} \sum_j C_{j\Omega} [D^{I*}_{MK}(\theta_i) \mid jK\rangle + (-1)^{I-j} D^{I*}_{M-K}(\theta_i) \mid j-K\rangle],$$

where the particle solutions are

$$\mid \Omega\rangle = \sum_j C_{j\Omega} \mid j\Omega\rangle.$$

The expansion coefficients $C_{j\Omega}$ and the particle eigenvalues are tabulated in Appendix D.

For odd-$A$ systems lacking axial symmetry $K$ and $\Omega$ are no longer good quantum numbers and must be summed over subject to the restriction

$$K - \Omega = 2n, \qquad n = 0, 1, 2, \ldots$$
$$\Omega > 0.$$

The only solutions available have been diagonalized in the symmetrized basis

$$\mid IMKj\Omega\rangle = \sqrt{\frac{2I+1}{16\pi^2}} [D^{I*}_{MK}(\theta_i) \mid j\Omega\rangle + (-1)^{I-j} D^{I*}_{M-K}(\theta_i) \mid j-\Omega\rangle].$$

## E. Odd–Odd Deformed Nuclei

In the strong coupling limit there are two possibilities of the projection quantum number

$$K_{1,2} = \Omega_{1,2} = \mid \Omega_n \pm \Omega_p \mid.$$

For independent particle motion the properly symmetrized state functions are

$$\mid EIMK\rangle = \sqrt{\frac{2I+1}{16\pi^2}} \sum_{j_n, j_p} C_{j_p\Omega_p} C_{j_n K \mp \Omega_p}$$
$$\times [D^{I*}_{MK}(\theta_i) \mid j_p\Omega_p\rangle \mid j_n K \mp \Omega_p\rangle$$
$$+ (-1)^{I-j_p-j_n} D^{I*}_{M-K}(\theta_i) \mid j_p - \Omega_p\rangle \mid j_n - K \pm \Omega_p\rangle].$$

For the special case $K = 0$

$$| EIM0\rangle = \sqrt{\frac{2I+1}{16\pi^2}} \sum_{j_n, j_p} C_{j_p\Omega_p} C_{j_n-\Omega_p} \times [| j_p\Omega_p\rangle | j_n - \Omega_p\rangle$$

$$+ (-1)^{I-j_p-j_n} | j_p - \Omega_p\rangle | j_n\Omega_p\rangle] D^{I*}_{M0}(\theta_i).$$

The particle solutions for this case, where different from the single particle solutions, are not available.

For the case of odd-odd nuclei without axial symmetry the problem is as yet unsolved.

APPENDIX D

# Deformed Single-Particle Solutions

In this appendix are given the eigenfunctions of the Hamiltonian and the eigenvalues to which they belong

$$H = \frac{p^2}{2m} + \frac{m}{2}[\omega^2(X_1{}^2 + X_2{}^2) + \omega_3{}^2 X_3{}^2] + \hbar\omega_0[\varkappa \mathbf{l} \cdot \mathbf{s} + \varkappa\mu \mathbf{l}^2]. \tag{1}$$

There are two representations with respect to which these solutions may be calculated: (1) the uncoupled basis $| Nl\Lambda\Sigma\rangle$ defined by (Eq. (4-17))

$$\begin{aligned}
H_{00} \mid Nl\Lambda\Sigma\rangle &= \hbar\omega_0(N + \tfrac{3}{2}) \mid Nl\Lambda\Sigma\rangle, \qquad N = 0, 1, 2, \ldots \\
\mathbf{l}^2 \mid Nl\Lambda\Sigma\rangle &= l(l+1) \mid Nl\Lambda\Sigma\rangle, \qquad l = N, N-2, \ldots, 1, \text{ or } 0 \\
l_3 \mid Nl\Lambda\Sigma\rangle &= \Lambda \mid Nl\Lambda\Sigma\rangle, \qquad |\Lambda| \leq l \\
s_3 \mid Nl\Lambda\Sigma\rangle &= \Sigma \mid Nl\Lambda\Sigma\rangle, \qquad \Sigma = \pm\tfrac{1}{2}
\end{aligned} \tag{2a}$$

so that the deformed particle solutions are

$$| N\Omega\rangle = \sum_{l\Lambda} a_{l\Lambda} \mid Nl\Lambda\Sigma\rangle, \tag{2b}$$

and (2) the coupled basis $| Nlj\Omega\rangle$ defined by

$$\begin{aligned}
H_{00} \mid Nlj\Omega\rangle &= \hbar\omega_0(N + \tfrac{3}{2}) \mid Nlj\Omega\rangle, \qquad N = 0, 1, 2, \ldots \\
\mathbf{l}^2 \mid Nlj\Omega\rangle &= l(l+1) \mid Nlj\Omega\rangle, \qquad l = N, N-2, \ldots, 1 \text{ or } 0 \\
\mathbf{j}^2 \mid Nlj\Omega\rangle &= j(j+1) \mid Nlj\Omega\rangle, \qquad j = l \pm \tfrac{1}{2} \\
j_3 \mid Nlj\Omega\rangle &= \Omega \mid Nlj\Omega\rangle, \qquad |\Omega| \leq j,
\end{aligned} \tag{3a}$$

so that the deformed solutions are

$$| N\Omega\rangle = \sum_{j} C_{j\Omega} \, | Nlj\Omega\rangle . \tag{3b}$$

Both representations are useful and are related through

$$C_{j\Omega} = \sum_{\Lambda} C(l\tfrac{1}{2}j;\ \Lambda\Sigma\Omega) a_{l\Lambda} . \tag{4}$$

The energy eigenvalues are then the solutions of

$$H \, | H\Omega\rangle = E_{\Omega} \, | N\Omega\rangle , \tag{5}$$

the $N$ label being appropriate since matrix elements of $H$ not diagonal in $N$ are ignored.

Table D-1 lists these energy eigenvalues, in units of $\hbar\omega_0$, and the normalized eigenvector components $C_{j\Omega}$ of the coupled representation (3) as a function of the deformation parameter $\beta$ for the oscillator shells $N \leq 7$. For each shell the values of the spin-orbit coupling parameter are given. The parameter $\varkappa$ has the value 0.05 throughout. For each $K = \frac{1}{2}$ state, the calculated values of the decoupling parameter $a$ are given. This parameter is defined most simply in the coupled representation as

$$a \equiv \sum_{j} (-1)^{j+1/2} (j + \tfrac{1}{2}) \, | C_{j1/2} |^2 . \tag{6}$$

Finally, for all states, the asymptotic quantum numbers $K\pi[Nn_3\Lambda]$ are listed.

The energy eigenvalues, in units of $\hbar\omega_0$, are presented graphically in Figs. D-1–D-5 for each of the shells. These figures are consistent with Table D-1 (pp. 190–224) in that the deformation parameter used is $\beta$.

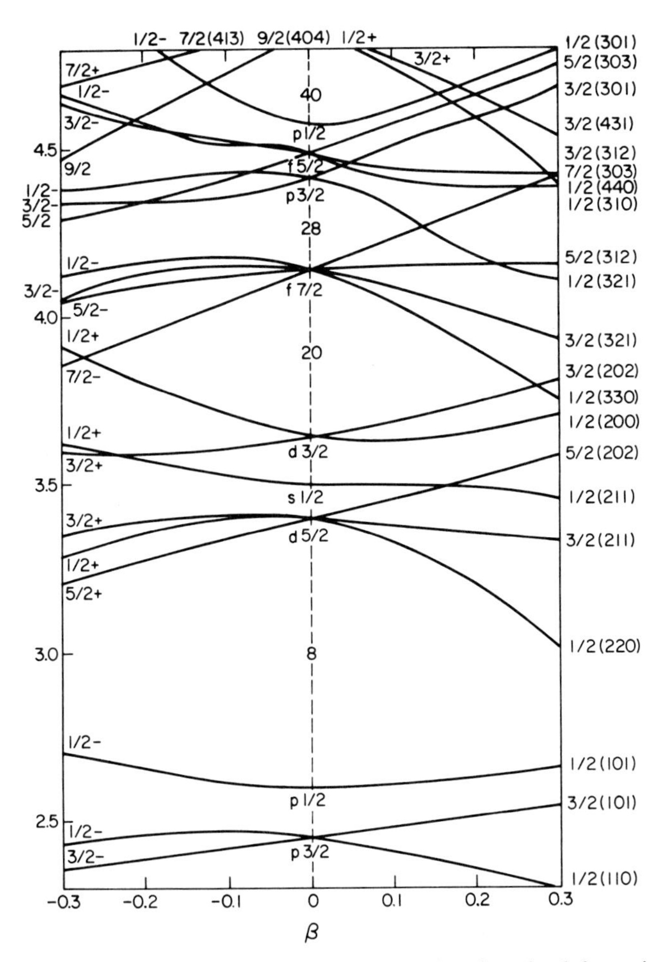


Fig. D-1. Energy eigenvalues in units of $\hbar\omega_0$ plotted against the deformation parameter $\beta$ for the Nilsson model in the oscillator shells $N = 1, 2, 3$. The spin orbit coupling parameters $\mu_N$ for these shells are $\mu_1 = \mu_2 = 0$, $\mu_3 = 0.35$. These energy levels are to be associated with odd neutron or proton nuclei in the $1p$, $2s$–$1d$ and $2p$–$1f$ shells.

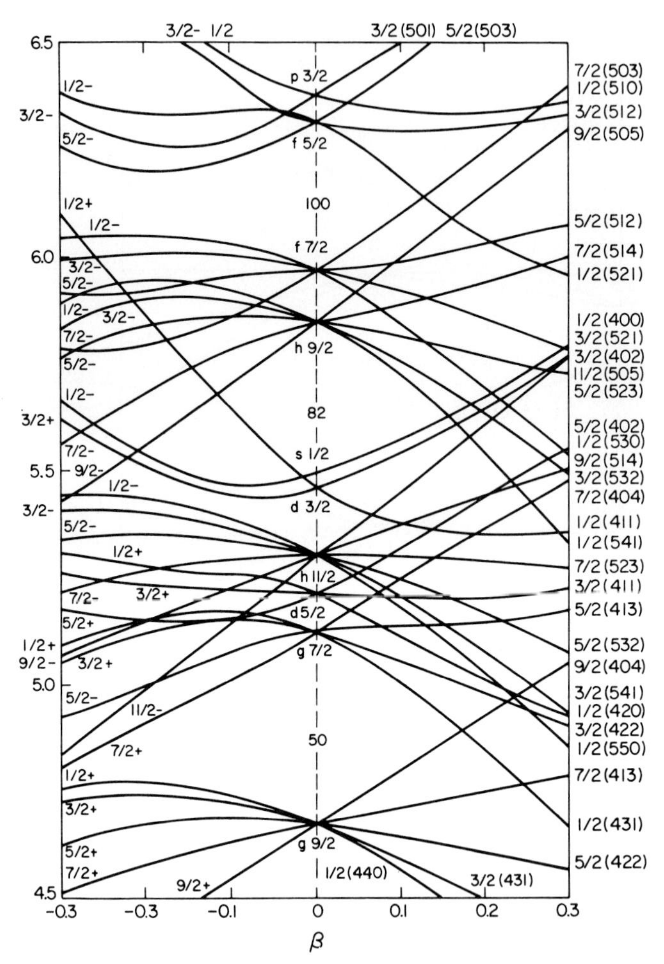


Fig. D-2. Energy eigenvalues in units of $\hbar\omega_0$ plotted against the deformation parameter $\beta$ for the Nilsson model in the oscillator shells $N = 4$ and 5. The spin orbit coupling parameters $\mu_N$ for these shells are $\mu_4 = 0.625$, $\mu_5 = 0.630$. These energy levels are to be associated with odd-$A$ nuclei in the range $50 < N$ or $Z < 82$.

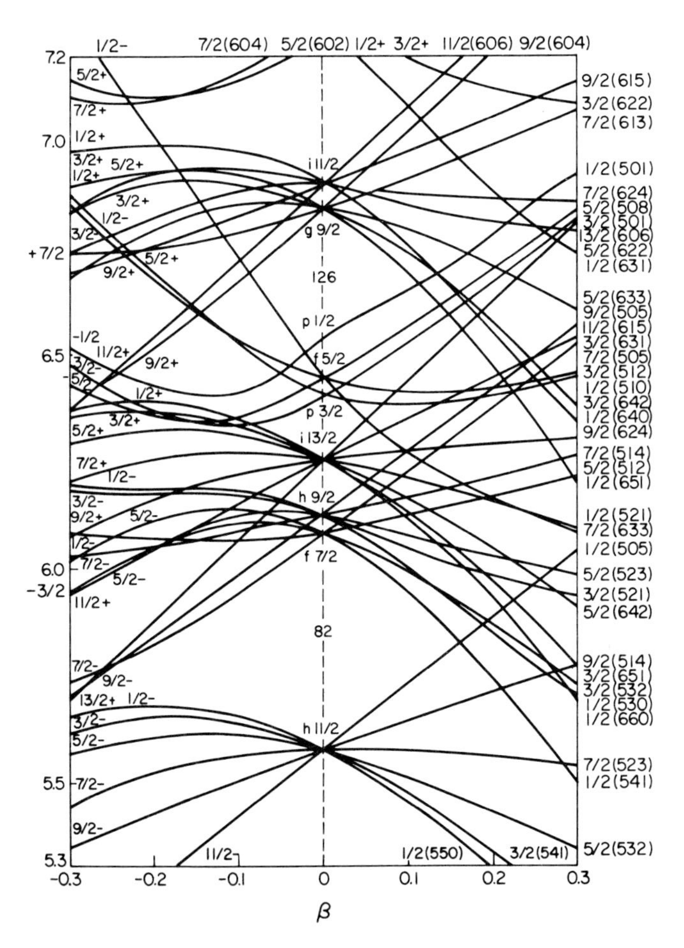

Fig. D-3. Energy eigenvalues in units of $\hbar\omega_0$ plotted against the deformation parameter $\beta$ for the Nilsson model in the oscillator shells $N = 5$ and 6. The spin orbit coupling parameters $\mu_N$ for these shells are $\mu_5 = 0.450$, $\mu_6 = 0.448$. These energy levels are to be associated with odd neutron nuclei in the range $82 < N < 126$.

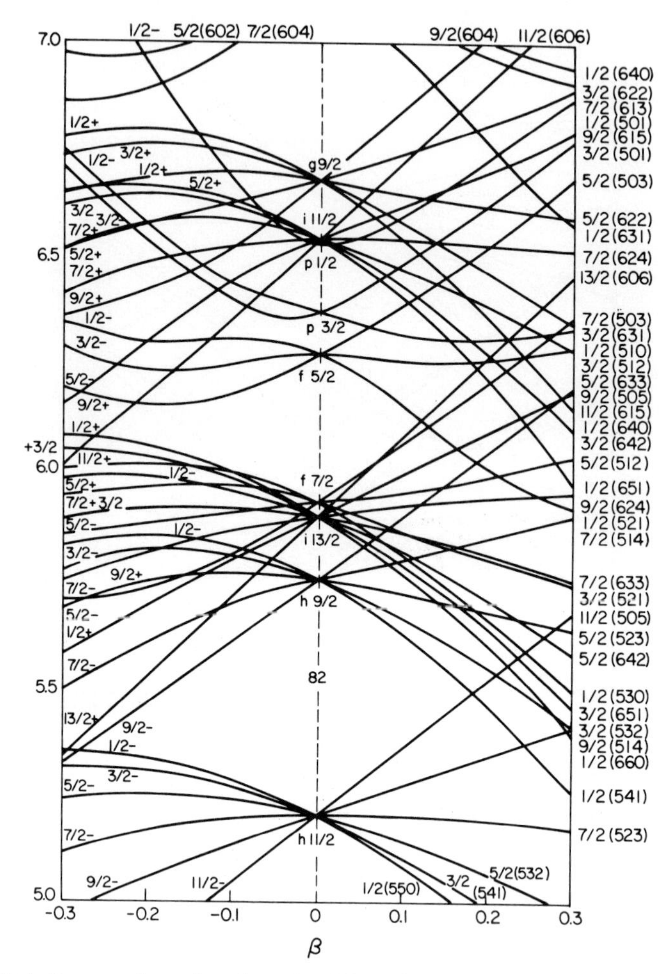


Fig. D-4. Energy eigenvalues in units of $\hbar\omega_0$ plotted against the deformation parameter $\beta$ for the Nilsson model in the oscillator shells $N = 5$ and 6. The spin orbit coupling parameters $\mu_N$ for these shells are $\mu_5 = 0.700$, $\mu_6 = 0.620$. These energy levels are to be associated with odd proton nuclei with $Z > 82$.

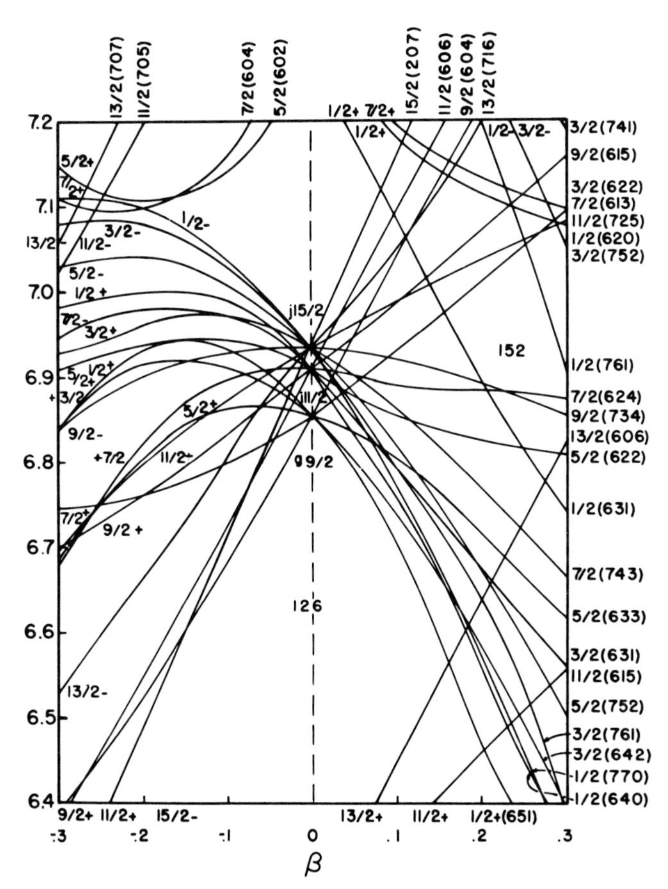


Fig. D-5. Energy eigenvalues in units of $\hbar\omega_0$ plotted against the deformation parameter $\beta$ for the Nilsson model in the oscillator shells $N = 6$ and 7. The spin orbit coupling parameters $\mu_N$ for these shells are $\mu_6 = 0.448$ and $\mu_7 = 0.434$. These energy levels are to be associated with odd neutron nuclei with $N > 126$.

| | $\beta =$ | -0.3 | -0.2 | -0.1 | 0.0 | 0.1 | 0.2 | 0.3 |
|---|---|---|---|---|---|---|---|---|

**N = 0 $\mu_0 = 0$ $\kappa = 0.05$**

**K = 1/2+[000]**

| | | -0.3 | -0.2 | -0.1 | 0.0 | 0.1 | 0.2 | 0.3 |
|---|---|---|---|---|---|---|---|---|
| E | | 1.500000 | 1.500000 | 1.500000 | 1.500000 | 1.500000 | 1.500000 | 1.500000 |
| a | | 1.000000 | 1.000000 | 1.000000 | 1.000000 | 1.000000 | 1.000000 | 1.000000 |
| L | J | | | | | | | |
| 0 | 1/2 | 1.000000 | 1.000000 | 1.000000 | 1.000000 | 1.000000 | 1.000000 | 1.000000 |

**N = 1 $\mu_1 = 0$ $\kappa = 0.05$**

**K = 3/2-[101]**

| | | -0.3 | -0.2 | -0.1 | 0.0 | 0.1 | 0.2 | 0.3 |
|---|---|---|---|---|---|---|---|---|
| E | | 2.355383 | 2.386922 | 2.418461 | 2.450000 | 2.481539 | 2.513078 | 2.544617 |
| L | J | | | | | | | |
| 1 | 3/2 | 1.000000 | 1.000000 | 1.000000 | 1.000000 | 1.000000 | 1.000000 | 1.000000 |

**K = 1/2-[110]**

| | | -0.3 | -0.2 | -0.1 | 0.0 | 0.1 | 0.2 | 0.3 |
|---|---|---|---|---|---|---|---|---|
| E | | 2.435664 | 2.457309 | 2.466623 | 2.450000 | 2.408094 | 2.354506 | 2.296406 |
| a | | -.803978 | -1.156971 | -1.698247 | -2.000000 | -1.846246 | -1.650080 | -1.512013 |
| L | J | | | | | | | |
| 1 | 1/2 | .631406 | .530103 | .317150 | .000000 | -.226387 | -.341526 | -.403314 |
| 1 | 3/2 | .775452 | .847933 | .948375 | 1.000000 | .974037 | .939872 | .915062 |

**K = 1/2-[101]**

| | | -0.3 | -0.2 | -0.1 | 0.0 | 0.1 | 0.2 | 0.3 |
|---|---|---|---|---|---|---|---|---|
| E | | 2.708953 | 2.655769 | 2.614916 | 2.600000 | 2.610367 | 2.632415 | 2.658976 |
| a | | -.196022 | .156971 | .698247 | 1.000000 | .846246 | .650080 | .512013 |
| L | J | | | | | | | |
| 1 | 1/2 | .775452 | .847933 | .948375 | 1.000000 | .974037 | .939872 | .915062 |
| 1 | 3/2 | -.631406 | -.530103 | -.317150 | .000000 | .226387 | .341526 | .403314 |

**N = 2 $\mu_2 = 0$ $\kappa = 0.05$**

**K = 5/2+[202]**

| | | -0.3 | -0.2 | -0.1 | 0.0 | 0.1 | 0.2 | 0.3 |
|---|---|---|---|---|---|---|---|---|
| E | | 3.210765 | 3.273843 | 3.336922 | 3.400000 | 3.463078 | 3.526157 | 3.589235 |
| L | J | | | | | | | |
| 2 | 5/2 | 1.000000 | 1.000000 | 1.000000 | 1.000000 | 1.000000 | 1.000000 | 1.000000 |

**K = 3/2+[211]**

| | | -0.3 | -0.2 | -0.1 | 0.0 | 0.1 | 0.2 | 0.3 |
|---|---|---|---|---|---|---|---|---|
| E | | 3.357362 | 3.391555 | 3.405467 | 3.400000 | 3.382784 | 3.359638 | 3.333443 |
| L | J | | | | | | | |
| 2 | 3/2 | .578305 | .406489 | .185596 | .000000 | -.120661 | -.196016 | -.245147 |
| 2 | 5/2 | .815821 | .913656 | .982626 | 1.000000 | .992694 | .980601 | .969486 |

| | $\beta$ = | −0.3 | −0.2 | −0.1 | 0.0 | 0.1 | 0.2 | 0.3 |
|---|---|---|---|---|---|---|---|---|
| **K = 3/2+[202]** | | | | | | | | |
| E | | 3.598020 | 3.595367 | 3.612994 | 3.650000 | 3.698755 | 3.753440 | 3.811175 |
| L | J | | | | | | | |
| 2 | 3/2 | .815821 | .913656 | .982626 | 1.000000 | .992694 | .980601 | .969486 |
| 2 | 5/2 | −.578305 | −.406489 | −.185596 | .000000 | .120661 | .196016 | .245147 |
| **K = 1/2+[220]** | | | | | | | | |
| E | | 3.295694 | 3.352608 | 3.400388 | 3.400000 | 3.318229 | 3.209946 | 3.093260 |
| a | | 1.562296 | 1.775203 | 2.263813 | 3.000000 | 2.623795 | 2.252148 | 1.996905 |
| L | J | | | | | | | |
| 0 | 1/2 | .777412 | .738537 | −.595822 | .000000 | .386178 | .490298 | .528947 |
| 2 | 3/2 | .213994 | .163659 | −.072361 | .000000 | −.124850 | −.231113 | −.297834 |
| 2 | 5/2 | −.591471 | −.654047 | .799850 | 1.000000 | .913936 | .840354 | .794676 |
| **K = 1/2+[211]** | | | | | | | | |
| E | | 3.621744 | 3.579517 | 3.530474 | 3.500000 | 3.506170 | 3.491019 | 3.462634 |
| a | | .929327 | 1.235668 | 1.391437 | 1.000000 | .836797 | .193252 | −.090199 |
| L | J | | | | | | | |
| 0 | 1/2 | .374241 | .510045 | .747861 | 1.000000 | .822034 | .553235 | .389818 |
| 2 | 3/2 | .598425 | .498807 | .313040 | .000000 | −.402920 | −.662512 | −.746496 |
| 2 | 5/2 | .708401 | .700747 | .585414 | .000000 | −.402387 | −.504984 | −.539245 |
| **K = 1/2+[200]** | | | | | | | | |
| E | | 3.916415 | 3.807110 | 3.713756 | 3.650000 | 3.630983 | 3.659801 | 3.710253 |
| a | | −.491622 | −1.010871 | −1.655250 | −2.000000 | −1.460592 | −.445401 | .093294 |
| L | J | | | | | | | |
| 0 | 1/2 | −.505544 | −.440927 | −.292746 | .000000 | .418481 | .673453 | .753828 |
| 2 | 3/2 | .772072 | .851121 | .946979 | 1.000000 | .906680 | .712506 | .595011 |
| 2 | 5/2 | −.385137 | −.284914 | −.132400 | .000000 | −.052968 | −.196969 | −.278756 |

**N = 3 $\mu_3$ = 0.35 $\kappa$ = 0.05**

| | | | | | | | | |
|---|---|---|---|---|---|---|---|---|
| **K = 7/2−[303]** | | | | | | | | |
| E | | 3.856148 | 3.950765 | 4.045383 | 4.140000 | 4.234617 | 4.329235 | 4.423852 |
| L | J | | | | | | | |
| 3 | 7/2 | 1.000000 | 1.000000 | 1.000000 | 1.000000 | 1.000000 | 1.000000 | 1.000000 |
| **K = 5/2−[312]** | | | | | | | | |
| E | | 4.049435 | 4.094196 | 4.122654 | 4.140000 | 4.150908 | 4.158158 | 4.163243 |
| L | J | | | | | | | |
| 3 | 5/2 | .449734 | .272721 | .114901 | .000000 | −.078552 | −.132846 | −.171655 |
| 3 | 7/2 | .893163 | .962093 | .993377 | 1.000000 | .996910 | .991137 | .985157 |
| **K = 5/2−[303]** | | | | | | | | |
| E | | 4.296712 | 4.346569 | 4.412729 | 4.490000 | 4.573709 | 4.661077 | 4.750609 |

| L | J | $\beta$ = -0.3 | -0.2 | -0.1 | 0.0 | 0.1 | 0.2 | 0.3 |
|---|---|---|---|---|---|---|---|---|
| 3 | 5/2 | .893163 | .962093 | .993377 | 1.000000 | .996910 | .991137 | .985157 |
| 3 | 7/2 | -.449734 | -.272721 | -.114901 | .000000 | .078552 | .132846 | .171655 |

**K = 3/2-[321]**

| | | -0.3 | -0.2 | -0.1 | 0.0 | 0.1 | 0.2 | 0.3 |
|---|---|---|---|---|---|---|---|---|
| E | | 4.058294 | 4.127055 | 4.159487 | 4.140000 | 4.086107 | 4.015130 | 3.935361 |
| L | J | | | | | | | |
| 1 | 3/2 | .739860 | -.596861 | -.299648 | .000000 | .171532 | .262184 | .313374 |
| 3 | 5/2 | .079959 | -.018094 | .035552 | .000000 | -.082845 | -.158316 | -.215799 |
| 3 | 7/2 | -.667993 | .802141 | .953387 | 1.000000 | .981689 | .951943 | .924785 |

**K = 3/2-[301]**

| | | -0.3 | -0.2 | -0.1 | 0.0 | 0.1 | 0.2 | 0.3 |
|---|---|---|---|---|---|---|---|---|
| E | | 4.346269 | 4.344188 | 4.363833 | 4.415000 | 4.515136 | 4.594744 | 4.682981 |
| L | J | | | | | | | |
| 1 | 3/2 | .511911 | .702096 | .910754 | 1.000000 | .734772 | .875280 | .898435 |
| 3 | 5/2 | .577347 | .472120 | .308284 | .000000 | .674539 | .454419 | .382804 |
| 3 | 7/2 | .630095 | .533069 | .274752 | .000000 | -.071463 | -.165496 | -.215118 |

**K = 3/2-[312]**

| | | -0.3 | -0.2 | -0.1 | 0.0 | 0.1 | 0.2 | 0.3 |
|---|---|---|---|---|---|---|---|---|
| E | | 4.640437 | 4.573757 | 4.521679 | 4.490000 | 4.443757 | 4.435126 | 4.426657 |
| L | J | | | | | | | |
| 1 | 3/2 | -.436525 | -.388352 | -.284146 | .000000 | .656267 | -.406380 | -.307589 |
| 3 | 5/2 | .812574 | .881348 | .950630 | 1.000000 | -.733576 | .876607 | .898272 |
| 3 | 7/2 | -.386224 | -.269086 | -.124756 | .000000 | -.176578 | .257712 | .313842 |

**K = 1/2-[330]**

| | | -0.3 | -0.2 | -0.1 | 0.0 | 0.1 | 0.2 | 0.3 |
|---|---|---|---|---|---|---|---|---|
| E | | 4.121973 | 4.170562 | 4.180494 | 4.140000 | 4.046901 | 3.912771 | 3.754219 |
| a | | -2.350736 | -3.174798 | -3.795419 | -4.000000 | -3.815425 | -3.382556 | -2.926407 |
| L | J | | | | | | | |
| 1 | 1/2 | .401664 | -.237282 | -.067564 | .000000 | -.050722 | -.142321 | -.217959 |
| 1 | 3/2 | .577011 | -.496866 | -.297681 | .000000 | .279574 | .453880 | .540891 |
| 3 | 5/2 | -.158885 | .084461 | .025432 | .000000 | -.046885 | -.121979 | -.189336 |
| 3 | 7/2 | -.693167 | .830474 | .951932 | 1.000000 | .957636 | .871125 | .789989 |

**K = 1/2-[321]**

| | | -0.3 | -0.2 | -0.1 | 0.0 | 0.1 | 0.2 | 0.3 |
|---|---|---|---|---|---|---|---|---|
| E | | 4.379312 | 4.402917 | 4.437763 | 4.415000 | 4.332125 | 4.221074 | 4.114343 |
| a | | .097011 | .430525 | -.446342 | -2.000000 | -.499078 | .495455 | .605673 |
| L | J | | | | | | | |
| 1 | 1/2 | .685994 | .733897 | .624698 | .000000 | -.417920 | -.503359 | .503304 |
| 1 | 3/2 | .020678 | .252410 | .662205 | 1.000000 | .732165 | .426708 | -.251867 |
| 3 | 5/2 | -.499030 | -.479595 | -.321911 | .000000 | -.471397 | -.639692 | .677414 |
| 3 | 7/2 | .529107 | .409479 | .260019 | .000000 | -.258964 | -.394136 | .473665 |

**K = 1/2-[310]**

| | | -0.3 | -0.2 | -0.1 | 0.0 | 0.1 | 0.2 | 0.3 |
|---|---|---|---|---|---|---|---|---|
| E | | 4.665707 | 4.579575 | 4.510109 | 4.490000 | 4.420566 | 4.394216 | 4.382664 |
| a | | -.205158 | -.049971 | 1.043300 | 3.000000 | 1.289938 | .016709 | -.417406 |
| L | J | | | | | | | |
| 1 | 1/2 | .122764 | .012052 | -.280639 | .000000 | .144891 | -.027196 | -.114857 |
| 1 | 3/2 | .612031 | .670981 | .576472 | .000000 | .560883 | .699427 | .699980 |

| L | J | $\beta$ = −0.3 | −0.2 | −0.1 | 0.0 | 0.1 | 0.2 | 0.3 |
|---|---|---|---|---|---|---|---|---|
| 3 | 5/2 | .651408 | .659964 | .754498 | 1.000000 | .806739 | .658419 | .601983 |
| 3 | 7/2 | .431295 | .337766 | .140194 | .000000 | −.116574 | −.276669 | −.366676 |

**K = 1/2 − [301]**

| L | J | −0.3 | −0.2 | −0.1 | 0.0 | 0.1 | 0.2 | 0.3 |
|---|---|---|---|---|---|---|---|---|
| E | | 5.010712 | 4.835416 | 4.670868 | 4.565000 | 4.621173 | 4.703469 | 4.791069 |
| a | | .458884 | .794243 | 1.198462 | 1.000000 | 1.024565 | .870393 | .738141 |
| 1 | 1/2 | −.594145 | .636355 | .725554 | 1.000000 | .895420 | .851842 | .828243 |
| 1 | 3/2 | .540415 | −.489078 | −.374900 | .000000 | .266802 | .350306 | .392464 |
| 3 | 5/2 | −.548992 | .572103 | .571366 | .000000 | −.353213 | −.377357 | −.377993 |
| 3 | 7/2 | .231411 | −.168977 | −.081004 | .000000 | −.047757 | −.096188 | −.130793 |

**N = 4 $\mu_4 = 0.625$ $\kappa = 0.05$**

**K = 9/2 + [404]**

| L | J | −0.3 | −0.2 | −0.1 | 0.0 | 0.1 | 0.2 | 0.3 |
|---|---|---|---|---|---|---|---|---|
| E | | 4.296530 | 4.422687 | 4.548843 | 4.675000 | 4.801157 | 4.927313 | 5.053470 |
| 4 | 9/2 | 1.000000 | 1.000000 | 1.000000 | 1.000000 | 1.000000 | 1.000000 | 1.000000 |

**K = 7/2 + [413]**

| L | J | −0.3 | −0.2 | −0.1 | 0.0 | 0.1 | 0.2 | 0.3 |
|---|---|---|---|---|---|---|---|---|
| E | | 4.518219 | 4.579635 | 4.630613 | 4.675000 | 4.715369 | 4.753240 | 4.789496 |
| 4 | 7/2 | .324698 | .186044 | .078270 | .000000 | −.056519 | −.098141 | −.129615 |
| 4 | 9/2 | .945818 | .982541 | .996932 | 1.000000 | .998402 | .995173 | .991564 |

**K = 7/2 + [404]**

| L | J | −0.3 | −0.2 | −0.1 | 0.0 | 0.1 | 0.2 | 0.3 |
|---|---|---|---|---|---|---|---|---|
| E | | 4.808694 | 4.904974 | 5.011691 | 5.125000 | 5.242327 | 5.362152 | 5.483591 |
| 4 | 7/2 | .945818 | .982541 | .996932 | 1.000000 | .998402 | .995173 | .991564 |
| 4 | 9/2 | −.324698 | −.186044 | −.078270 | .000000 | .056519 | .098141 | .129615 |

**K = 5/2 + [422]**

| L | J | −0.3 | −0.2 | −0.1 | 0.0 | 0.1 | 0.2 | 0.3 |
|---|---|---|---|---|---|---|---|---|
| E | | 4.629043 | 4.673988 | 4.686680 | 4.675000 | 4.646915 | 4.608133 | 4.562402 |
| 2 | 5/2 | −.508110 | −.310267 | −.131347 | .000000 | .090247 | .151955 | .194952 |
| 4 | 7/2 | .055702 | .081002 | .056209 | .000000 | −.062161 | −.117537 | −.162999 |
| 4 | 9/2 | .859489 | .947192 | .989742 | 1.000000 | .993978 | .981374 | .967174 |

**K = 5/2 + [413]**

| L | J | −0.3 | −0.2 | −0.1 | 0.0 | 0.1 | 0.2 | 0.3 |
|---|---|---|---|---|---|---|---|---|
| E | | 4.923183 | 4.994990 | 5.080427 | 5.125000 | 5.138841 | 5.155257 | 5.175472 |
| 2 | 5/2 | .688293 | .767345 | .646377 | .000000 | −.152520 | −.172558 | −.167549 |
| 4 | 7/2 | .626148 | .609498 | .761833 | 1.000000 | .985414 | .974500 | .966065 |
| 4 | 9/2 | .366323 | .199232 | .042514 | .000000 | .075473 | .143433 | .196585 |

| L | J | β = −0.3 | −0.2 | −0.1 | 0.0 | 0.1 | 0.2 | 0.3 |
|---|---|---|---|---|---|---|---|---|
| **K = 5/2+[402]** | | | | | | | | |
| E | | 5.176422 | 5.154287 | 5.150776 | 5.212500 | 5.321362 | 5.438345 | 5.558478 |
| 2 | 5/2 | −.517762 | −.561173 | .751628 | 1.000000 | .984171 | .973208 | .966396 |
| 4 | 7/2 | .777712 | .788639 | −.645330 | .000000 | .156413 | .191139 | .200374 |
| 4 | 9/2 | −.356491 | −.251264 | .136397 | .000000 | −.079450 | −.127798 | −.161026 |
| **K = 3/2+[431]** | | | | | | | | |
| E | | 4.722666 | 4.739122 | 4.722994 | 4.675000 | 4.598392 | 4.498664 | 4.381870 |
| 2 | 3/2 | −.195906 | −.082788 | −.018597 | .000000 | −.013070 | −.041631 | −.073570 |
| 2 | 5/2 | −.452479 | −.326014 | −.165851 | .000000 | .145433 | .258188 | .339291 |
| 4 | 7/2 | .107651 | .066307 | .035817 | .000000 | −.048951 | −.104717 | −.157521 |
| 4 | 9/2 | .863304 | .939396 | .985325 | 1.000000 | .988070 | .959500 | .924476 |
| **K = 3/2+[422]** | | | | | | | | |
| E | | 5.052422 | 5.122284 | 5.152334 | 5.125000 | 5.058965 | 4.979796 | 4.901742 |
| 2 | 3/2 | −.728269 | −.609903 | −.259571 | .000000 | .172574 | .259838 | .302309 |
| 2 | 5/2 | −.063097 | −.087233 | .117928 | .000000 | −.199871 | −.235515 | −.212908 |
| 4 | 7/2 | .624008 | .775343 | .958290 | 1.000000 | .961238 | .919986 | .893605 |
| 4 | 9/2 | −.276146 | −.138752 | −.019884 | .000000 | .079324 | .175052 | .254458 |
| **K = 3/2+[411]** | | | | | | | | |
| E | | 5.265591 | 5.232072 | 5.226148 | 5.212500 | 5.202659 | 5.209863 | 5.226002 |
| 2 | 3/2 | .297693 | .358006 | .338317 | .000000 | −.121742 | −.158170 | −.181051 |
| 2 | 5/2 | .678642 | .787479 | .926605 | 1.000000 | .956271 | .909112 | .876839 |
| 4 | 7/2 | .571802 | .419469 | −.019006 | .000000 | .231496 | .318640 | .348971 |
| 4 | 9/2 | .351946 | .275235 | .163043 | .000000 | −.130894 | −.216716 | −.276755 |
| **K = 3/2+[402]** | | | | | | | | |
| E | | 5.623557 | 5.507679 | 5.436602 | 5.462500 | 5.551906 | 5.660520 | 5.776151 |
| 2 | 3/2 | −.585341 | .702137 | .904333 | 1.000000 | .977357 | .951700 | .932962 |
| 2 | 5/2 | .575088 | −.515734 | −.316211 | .000000 | .156352 | .226688 | .265904 |
| 4 | 7/2 | −.521601 | .467432 | .282905 | .000000 | −.141547 | −.202803 | −.234256 |
| 4 | 9/2 | .233631 | −.150099 | −.046441 | .000000 | −.017098 | −.041839 | −.063259 |
| **K = 1/2+[440]** | | | | | | | | |
| E | | 4.759739 | 4.769370 | 4.740771 | 4.675000 | 4.572814 | 4.435521 | 4.266325 |
| a | | 4.300502 | 4.708481 | 4.929671 | 5.000000 | 4.929701 | 4.709590 | 4.335115 |
| 0 | 1/2 | .239462 | .110521 | .027217 | .000000 | .023345 | .080993 | .151182 |
| 2 | 3/2 | .040051 | .002126 | −.003744 | .000000 | −.010300 | −.047054 | −.105789 |
| 2 | 5/2 | −.478333 | −.347455 | −.181769 | .000000 | .178791 | .334406 | .451313 |
| 4 | 7/2 | −.012005 | .011439 | .011484 | .000000 | −.019561 | −.052722 | −.098745 |
| 4 | 9/2 | .843865 | .931088 | .982890 | 1.000000 | .983362 | .936279 | .867478 |

| | $\beta$ = | -0.3 | -0.2 | -0.1 | 0.0 | 0.1 | 0.2 | 0.3 |
|---|---|---|---|---|---|---|---|---|

K = 1/2+[431]

| | | -0.3 | -0.2 | -0.1 | 0.0 | 0.1 | 0.2 | 0.3 |
|---|---|---|---|---|---|---|---|---|
| | E | 5.088871 | 5.152975 | 5.176959 | 5.125000 | 5.012779 | 4.849937 | 4.671607 |
| | a | -1.015245 | -1.847523 | -3.636846 | -4.000000 | -3.578336 | -2.624253 | -2.065000 |
| L | J | | | | | | | |
| 0 | 1/2 | .525797 | .438932 | -.129117 | .000000 | -.093368 | -.219962 | -.260242 |
| 2 | 3/2 | .518204 | .459094 | -.283600 | .000000 | .274598 | .446478 | .518446 |
| 2 | 5/2 | -.200085 | -.283838 | .129178 | .000000 | -.169229 | -.253788 | -.198194 |
| 4 | 7/2 | -.572478 | -.702407 | .941263 | 1.000000 | .940352 | .810110 | .732881 |
| 4 | 9/2 | -.295359 | -.150440 | .015387 | .000000 | .054567 | .177728 | .295114 |

K = 1/2+[420]

| | | -0.3 | -0.2 | -0.1 | 0.0 | 0.1 | 0.2 | 0.3 |
|---|---|---|---|---|---|---|---|---|
| | E | 5.309775 | 5.274648 | 5.258102 | 5.212500 | 5.113503 | 5.014164 | 4.928324 |
| | a | -.117298 | .518579 | 2.482763 | 3.000000 | 2.461428 | 1.446702 | 1.056941 |
| L | J | | | | | | | |
| 0 | 1/2 | .498238 | .509196 | -.418179 | .000000 | .295125 | .387751 | .423203 |
| 2 | 3/2 | -.199077 | -.116883 | -.002536 | .000000 | -.098019 | -.075222 | -.055354 |
| 2 | 5/2 | -.435780 | -.591644 | .872802 | 1.000000 | .906479 | .753218 | .652277 |
| 4 | 7/2 | .620672 | .542482 | -.180773 | .000000 | .230823 | .444424 | .511455 |
| 4 | 9/2 | -.370121 | -.287625 | .175092 | .000000 | -.168254 | -.281320 | -.361639 |

K = 1/2+[411]

| | | -0.3 | -0.2 | -0.1 | 0.0 | 0.1 | 0.2 | 0.3 |
|---|---|---|---|---|---|---|---|---|
| | E | 6.095153 | 5.864637 | 5.650407 | 5.462500 | 5.382689 | 5.359034 | 5.361693 |
| | a | -.170408 | -.316372 | -.443420 | -2.000000 | -1.230749 | -.973716 | -.819513 |
| L | J | | | | | | | |
| 0 | 1/2 | -.558322 | -.587330 | -.643822 | .000000 | -.404377 | -.331972 | -.274309 |
| 2 | 3/2 | .638113 | .668571 | .698383 | 1.000000 | .839536 | .760180 | .712822 |
| 2 | 5/2 | -.436933 | -.380224 | -.267270 | .000000 | .275201 | .418319 | .472643 |
| 4 | 7/2 | .277163 | .240021 | .159289 | .000000 | -.233677 | -.355559 | -.410276 |
| 4 | 9/2 | -.115578 | -.076648 | -.030801 | .000000 | -.036291 | -.102509 | -.157863 |

K = 1/2+[400]

| | | -0.3 | -0.2 | -0.1 | 0.0 | 0.1 | 0.2 | 0.3 |
|---|---|---|---|---|---|---|---|---|
| | E | 5.667637 | 5.544153 | 5.464154 | 5.500000 | 5.577824 | 5.685561 | 5.800876 |
| | a | .002449 | -.063164 | -.332189 | 1.000000 | .417956 | .441676 | .492458 |
| L | J | | | | | | | |
| 0 | 1/2 | .325930 | .436906 | .627063 | 1.000000 | .860301 | .827345 | .809365 |
| 2 | 3/2 | .532015 | .573218 | .657124 | .000000 | .458324 | .463586 | .456994 |
| 2 | 5/2 | .591904 | .551450 | .342135 | .000000 | -.204828 | -.285370 | -.328905 |
| 4 | 7/2 | .458337 | .393188 | .236306 | .000000 | -.086434 | -.130414 | -.152387 |
| 4 | 9/2 | .224293 | .147784 | .045650 | .000000 | .019899 | .046309 | .068445 |

N = 5 $\mu_s$ = 0.45 $\kappa$ = 0.05 odd neutron nuclei

K = 11/2-[505]

| | | -0.3 | -0.2 | -0.1 | 0.0 | 0.1 | 0.2 | 0.3 |
|---|---|---|---|---|---|---|---|---|
| | E | 5.101913 | 5.259609 | 5.417304 | 5.575000 | 5.732696 | 5.890392 | 6.048087 |
| L | J | | | | | | | |
| 5 | 11/2 | 1.000000 | 1.000000 | 1.000000 | 1.000000 | 1.000000 | 1.000000 | 1.000000 |

**K = 9/2-[514]**

| β = | | -0.3 | -0.2 | -0.1 | 0.0 | 0.1 | 0.2 | 0.3 |
|---|---|---|---|---|---|---|---|---|
| E | | 5.340230 | 5.424288 | 5.501760 | 5.575000 | 5.645503 | 5.714186 | 5.781618 |
| L | J | | | | | | | |
| 5 | 9/2 | .235020 | .133933 | .057273 | .000000 | -.043232 | -.076504 | -.102661 |
| 5 | 11/2 | .971991 | .990990 | .998359 | 1.000000 | .999065 | .997069 | .994716 |

**K = 9/2-[505]**

| β = | | -0.3 | -0.2 | -0.1 | 0.0 | 0.1 | 0.2 | 0.3 |
|---|---|---|---|---|---|---|---|---|
| E | | 5.697448 | 5.834164 | 5.977466 | 6.125000 | 6.275271 | 6.427363 | 6.580704 |
| L | J | | | | | | | |
| 5 | 9/2 | .971991 | .990990 | .998359 | 1.000000 | .999065 | .997069 | .994716 |
| 5 | 11/2 | -.235020 | -.133933 | -.057273 | .000000 | .043232 | .076504 | .102661 |

**K = 7/2-[523]**

| β = | | -0.3 | -0.2 | -0.1 | 0.0 | 0.1 | 0.2 | 0.3 |
|---|---|---|---|---|---|---|---|---|
| E | | 5.447537 | 5.525285 | 5.563313 | 5.575000 | 5.571834 | 5.560017 | 5.542831 |
| L | J | | | | | | | |
| 3 | 7/2 | -.575550 | -.345011 | -.136850 | .000000 | .084996 | .139733 | .176782 |
| 5 | 9/2 | .048688 | .071216 | .048074 | .000000 | -.050426 | -.094884 | -.131966 |
| 5 | 11/2 | .816316 | .935893 | .989425 | 1.000000 | .995104 | .985633 | .975363 |

**K = 7/2-[503]**

| β = | | -0.3 | -0.2 | -0.1 | 0.0 | 0.1 | 0.2 | 0.3 |
|---|---|---|---|---|---|---|---|---|
| E | | 5.736133 | 5.825662 | 5.944570 | 6.080000 | 6.230766 | 6.380058 | 6.532577 |
| L | J | | | | | | | |
| 3 | 7/2 | .753677 | .898043 | .968820 | 1.000000 | .920436 | .956162 | .961269 |
| 5 | 9/2 | .418966 | .314984 | .214753 | .000000 | .386410 | .271633 | .236041 |
| 5 | 11/2 | .506397 | .307090 | .123566 | .000000 | -.059038 | -.109406 | -.142291 |

**K = 7/2-[514]**

| β = | | -0.3 | -0.2 | -0.1 | 0.0 | 0.1 | 0.2 | 0.3 |
|---|---|---|---|---|---|---|---|---|
| E | | 6.028625 | 6.050584 | 6.082881 | 6.125000 | 6.166635 | 6.218395 | 6.272296 |
| L | J | | | | | | | |
| 3 | 7/2 | -.317353 | -.272922 | -.206542 | .000000 | -.381541 | -.257350 | -.211448 |
| 5 | 9/2 | .906696 | .946421 | .975484 | 1.000000 | .920948 | .957712 | .962741 |
| 5 | 11/2 | -.277831 | -.172628 | -.075964 | .000000 | .079258 | .128680 | .168583 |

**K = 5/2-[532]**

| β = | | -0.3 | -0.2 | -0.1 | 0.0 | 0.1 | 0.2 | 0.3 |
|---|---|---|---|---|---|---|---|---|
| E | | 5.567792 | 5.605702 | 5.607978 | 5.575000 | 5.513421 | 5.431863 | 5.337159 |
| L | J | | | | | | | |
| 3 | 5/2 | -.174597 | -.071935 | -.015431 | .000000 | -.009426 | -.028582 | -.049191 |
| 3 | 7/2 | -.499391 | -.359439 | -.178243 | .000000 | .141294 | .241417 | .310087 |
| 5 | 9/2 | .090506 | .062438 | .036434 | .000000 | -.046262 | -.093968 | -.137110 |
| 5 | 11/2 | .843761 | .928294 | .983191 | 1.000000 | .988841 | .965438 | .939483 |

**K = 5/2-[512]**

| β = | | -0.3 | -0.2 | -0.1 | 0.0 | 0.1 | 0.2 | 0.3 |
|---|---|---|---|---|---|---|---|---|
| E | | 6.083202 | 6.072985 | 6.063106 | 6.080000 | 6.127218 | 6.170022 | 6.221361 |
| L | J | | | | | | | |
| 3 | 5/2 | .087436 | -.276801 | .127424 | .000000 | .004643 | -.056945 | -.095264 |
| 3 | 7/2 | .734181 | .615157 | .954025 | 1.000000 | .786050 | .880763 | .886985 |

| L | J (β =) | −0.3 | −0.2 | −0.1 | 0.0 | 0.1 | 0.2 | 0.3 |
|---|---|---|---|---|---|---|---|---|
| 5 | 9/2 | .545934 | .718753 | .213773 | .000000 | .612463 | .434443 | .381560 |
| 5 | 11/2 | .394068 | .168397 | .167034 | .000000 | -.083620 | -.179644 | -.242062 |

**K = 5/2 - [523]**

| | | −0.3 | −0.2 | −0.1 | 0.0 | 0.1 | 0.2 | 0.3 |
|---|---|---|---|---|---|---|---|---|
| E | | 5.939928 | 6.045211 | 6.129880 | 6.125000 | 6.071821 | 6.029875 | 5.987422 |
| L | J | | | | | | | |
| 3 | 5/2 | .811155 | .662999 | -.362125 | .000000 | .165391 | .235259 | .268566 |
| 3 | 7/2 | .115896 | .599080 | -.145814 | .000000 | -.593888 | -.378521 | -.287902 |
| 5 | 9/2 | -.494751 | -.328973 | .918275 | 1.000000 | .777726 | .875482 | .887713 |
| 5 | 11/2 | .289514 | .305470 | -.066146 | .000000 | .122822 | .186831 | .238642 |

**K = 5/2 - [503]**

| | | −0.3 | −0.2 | −0.1 | 0.0 | 0.1 | 0.2 | 0.3 |
|---|---|---|---|---|---|---|---|---|
| E | | 6.429844 | 6.359946 | 6.345958 | 6.430000 | 6.560618 | 6.704396 | 6.853293 |
| L | J | | | | | | | |
| 3 | 5/2 | .551270 | .691838 | .923250 | 1.000000 | .986172 | .969842 | .957276 |
| 3 | 7/2 | -.445146 | -.365360 | -.191843 | .000000 | .097250 | .150649 | .184975 |
| 5 | 9/2 | .670065 | .609321 | .331278 | .000000 | -.133758 | -.189630 | -.218125 |
| 5 | 11/2 | -.221267 | -.128841 | -.032565 | .000000 | -.010753 | -.027416 | -.042764 |

**K = 3/2 - [541]**

| | | −0.3 | −0.2 | −0.1 | 0.0 | 0.1 | 0.2 | 0.3 |
|---|---|---|---|---|---|---|---|---|
| E | | 5.623813 | 5.654434 | 5.636998 | 5.575000 | 5.472340 | 5.335598 | 5.173348 |
| L | J | | | | | | | |
| 1 | 3/2 | .300584 | .131211 | .029367 | .000000 | .019514 | .059463 | .100501 |
| 3 | 5/2 | -.005296 | -.021361 | -.009406 | .000000 | -.012041 | -.044076 | -.084682 |
| 3 | 7/2 | -.526060 | -.388629 | -.202330 | .000000 | .185391 | .329595 | .428989 |
| 5 | 9/2 | .016522 | .030690 | .022109 | .000000 | -.032963 | -.074258 | -.117190 |
| 5 | 11/2 | .795368 | .911237 | .978582 | 1.000000 | .981844 | .938282 | .885982 |

**K = 3/2 - [532]**

| | | −0.3 | −0.2 | −0.1 | 0.0 | 0.1 | 0.2 | 0.3 |
|---|---|---|---|---|---|---|---|---|
| E | | 5.953174 | 6.045425 | 6.107308 | 6.080000 | 5.995064 | 5.866483 | 5.729491 |
| L | J | | | | | | | |
| 1 | 3/2 | .709321 | .697291 | -.422063 | .000000 | .187163 | -.205009 | -.197721 |
| 3 | 5/2 | .389940 | .271030 | -.051549 | .000000 | -.226642 | .385294 | .452932 |
| 3 | 7/2 | -.227760 | -.493253 | .854804 | 1.000000 | .754181 | -.458287 | -.305579 |
| 5 | 9/2 | -.354776 | -.333270 | .234095 | .000000 | -.562724 | .732774 | .751253 |
| 5 | 11/2 | -.408741 | -.293191 | .183619 | .000000 | -.167796 | .250070 | .313049 |

**K = 3/2 - [521]**

| | | −0.3 | −0.2 | −0.1 | 0.0 | 0.1 | 0.2 | 0.3 |
|---|---|---|---|---|---|---|---|---|
| E | | 6.181173 | 6.177939 | 6.170761 | 6.125000 | 6.045684 | 5.987062 | 5.939609 |
| L | J | | | | | | | |
| 1 | 3/2 | .323785 | .194439 | .022162 | .000000 | .116401 | .248171 | .315069 |
| 3 | 5/2 | -.499671 | -.476879 | -.330527 | .000000 | .126649 | .056957 | .005564 |
| 3 | 7/2 | -.292686 | -.361364 | -.241932 | .000000 | .580852 | .739930 | .731820 |
| 5 | 9/2 | .669934 | .745983 | .908952 | 1.000000 | .791193 | .579743 | .511384 |
| 5 | 11/2 | -.333191 | -.218417 | -.074399 | .000000 | -.083874 | -.227089 | -.321911 |

**K = 3/2 - [501]**

| | | −0.3 | −0.2 | −0.1 | 0.0 | 0.1 | 0.2 | 0.3 |
|---|---|---|---|---|---|---|---|---|
| E | | 6.472154 | 6.377016 | 6.336823 | 6.405000 | 6.538078 | 6.680647 | 6.829176 |

| L | J | β = −0.3 | −0.2 | −0.1 | 0.0 | 0.1 | 0.2 | 0.3 |
|---|---|---|---|---|---|---|---|---|
| 1 | 3/2 | .398444 | .554497 | .818355 | 1.000000 | .872774 | .879450 | .876088 |
| 3 | 5/2 | .363587 | .360467 | .389819 | .000000 | .456245 | .401894 | .375910 |
| 3 | 7/2 | .635393 | .595590 | .356827 | .000000 | −.145987 | −.224374 | −.269451 |
| 5 | 9/2 | .486504 | .423708 | .220701 | .000000 | −.092914 | −.116764 | −.126611 |
| 5 | 11/2 | .261987 | .168347 | .047979 | .000000 | .012695 | .032720 | .050271 |

**K = 3/2 - [512]**

| L | J | −0.3 | −0.2 | −0.1 | 0.0 | 0.1 | 0.2 | 0.3 |
|---|---|---|---|---|---|---|---|---|
| E | | 6.857774 | 6.675577 | 6.520806 | 6.430000 | 6.406139 | 6.429818 | 6.470289 |
| 1 | 3/2 | −.378045 | −.388968 | −.388336 | .000000 | −.435093 | −.345563 | −.289840 |
| 3 | 5/2 | .682680 | .754146 | .857932 | 1.000000 | .851052 | .827552 | .803954 |
| 3 | 7/2 | −.426612 | −.346926 | −.206196 | .000000 | .195291 | .288964 | .338268 |
| 5 | 9/2 | .434008 | .389835 | .264211 | .000000 | −.218255 | −.328317 | −.379924 |
| 5 | 11/2 | −.143763 | −.087401 | −.028702 | .000000 | −.025118 | −.066715 | −.104322 |

**K = 1/2 - [550]**

| L | J | −0.3 | −0.2 | −0.1 | 0.0 | 0.1 | 0.2 | 0.3 |
|---|---|---|---|---|---|---|---|---|
| E | | 5.660770 | 5.679650 | 5.651349 | 5.575000 | 5.450873 | 5.280454 | 5.068094 |
| a | | −5.043315 | −5.599449 | −5.903167 | −6.000000 | −5.904252 | −5.613915 | −5.139005 |
| 1 | 1/2 | .124866 | .040565 | .005150 | .000000 | −.004335 | −.028561 | −.073650 |
| 1 | 3/2 | .257047 | .132105 | .035681 | .000000 | .034055 | .119948 | .222486 |
| 3 | 5/2 | −.056523 | −.018567 | −.003937 | .000000 | −.006388 | −.032731 | −.081631 |
| 3 | 7/2 | −.522132 | −.396212 | −.213061 | .000000 | .210914 | .388197 | .510056 |
| 5 | 9/2 | .029004 | .013993 | .007539 | .000000 | −.012212 | −.032738 | −.063772 |
| 5 | 11/2 | .801048 | .907402 | .976337 | 1.000000 | .976804 | .912117 | .821089 |

**K = 1/2 - [541]**

| L | J | −0.3 | −0.2 | −0.1 | 0.0 | 0.1 | 0.2 | 0.3 |
|---|---|---|---|---|---|---|---|---|
| E | | 6.023408 | 6.116620 | 6.141875 | 6.080000 | 5.946479 | 5.739759 | 5.503486 |
| a | | .750711 | −.421339 | −3.689234 | −4.000000 | −2.897391 | .134419 | .993775 |
| 1 | 1/2 | .594557 | .474626 | −.122708 | .000000 | −.103056 | −.257508 | −.323743 |
| 1 | 3/2 | .357132 | .466910 | −.391697 | .000000 | .353612 | .423352 | .382762 |
| 3 | 5/2 | −.487683 | −.380508 | .040193 | .000000 | −.165548 | −.426147 | −.526942 |
| 3 | 7/2 | −.135720 | −.440656 | .886562 | 1.000000 | .848677 | .471908 | .233474 |
| 5 | 9/2 | .380035 | .361602 | .021606 | .000000 | −.276494 | −.511107 | −.527950 |
| 5 | 11/2 | −.343914 | −.294966 | .208427 | .000000 | −.200573 | −.298217 | −.371179 |

**K = 1/2 - [530]**

| L | J | −0.3 | −0.2 | −0.1 | 0.0 | 0.1 | 0.2 | 0.3 |
|---|---|---|---|---|---|---|---|---|
| E | | 6.198505 | 6.185100 | 6.183796 | 6.125000 | 5.992596 | 5.839950 | 5.695436 |
| a | | .142756 | 1.401813 | 4.695198 | 5.000000 | 3.898769 | .816709 | −.215205 |
| 1 | 1/2 | −.070451 | .066709 | .125203 | .000000 | .048635 | −.004018 | −.068679 |
| 1 | 3/2 | −.478933 | 341909 | .010315 | .000000 | .087334 | .334395 | .449509 |
| 3 | 5/2 | −.359779 | 396537 | −.345368 | .000000 | .285878 | .284828 | .247515 |
| 3 | 7/2 | .396269 | .456745 | .016306 | .000000 | .305652 | .538626 | .494429 |
| 5 | 9/2 | .580868 | .678946 | .929858 | 1.000000 | .900975 | .678161 | .595685 |
| 5 | 11/2 | .376541 | .227646 | −.006052 | .000000 | −.055692 | −.238778 | −.364225 |

**K = 1/2 - [510]**

| L | J | −0.3 | −0.2 | −0.1 | 0.0 | 0.1 | 0.2 | 0.3 |
|---|---|---|---|---|---|---|---|---|
| E | | 6.880936 | 6.687638 | 6.520599 | 6.405000 | 6.392215 | 6.413435 | 6.453187 |
| a | | −.042947 | −.001407 | .175931 | −2.000000 | .191235 | −.146153 | −.317209 |

| L | J | β = −0.3 | −0.2 | −0.1 | 0.0 | 0.1 | 0.2 | 0.3 |
|---|---|---|---|---|---|---|---|---|
| 1 | 1/2 | .088225 | .048837 | −.034218 | .000000 | .008463 | −.053756 | −.091520 |
| 1 | 3/2 | .541658 | .617618 | .697587 | 1.000000 | .709476 | .680620 | .641688 |
| 3 | 5/2 | .552227 | .571011 | .626020 | .000000 | .638066 | .573012 | .543907 |
| 3 | 7/2 | .482248 | .417223 | .261933 | .000000 | −.226877 | −.361389 | −.429648 |
| 5 | 9/2 | .371639 | .326615 | .225005 | .000000 | −.193109 | −.263604 | −.293164 |
| 5 | 11/2 | .152280 | .096725 | .032634 | .000000 | .026049 | .073720 | .116116 |

K = 1/2 - [521]

| | | −0.3 | −0.2 | −0.1 | 0.0 | 0.1 | 0.2 | 0.3 |
|---|---|---|---|---|---|---|---|---|
| E | | 7.309260 | 7.014596 | 6.735857 | 6.430000 | 6.264103 | 6.160770 | 6.094165 |
| a | | .529376 | .737042 | .972254 | 3.000000 | .746609 | .945604 | .905866 |
| L | J | | | | | | | |
| 1 | 1/2 | −.532664 | −.579321 | .684480 | .000000 | −.473244 | −.508434 | .500717 |
| 1 | 3/2 | .526740 | .507807 | −.447896 | .000000 | .516463 | .304393 | −.173215 |
| 3 | 5/2 | −.555561 | −.566241 | .548042 | 1.000000 | −.593202 | −.495173 | .433455 |
| 3 | 7/2 | .284010 | .224920 | −.127638 | .000000 | −.291909 | −.434983 | .478122 |
| 5 | 9/2 | −.211979 | −.183263 | .118507 | .000000 | .265353 | .445203 | −.511919 |
| 5 | 11/2 | .067601 | .041419 | −.013801 | .000000 | .042361 | .127389 | −.201824 |

K = 1/2 - [501]

| | | −0.3 | −0.2 | −0.1 | 0.0 | 0.1 | 0.2 | 0.3 |
|---|---|---|---|---|---|---|---|---|
| E | | 6.516383 | 6.432571 | 6.409611 | 6.555000 | 6.650647 | 6.789458 | 6.936370 |
| a | | .663419 | .883341 | .749017 | 1.000000 | .965031 | .863335 | .771779 |
| L | J | | | | | | | |
| 1 | 1/2 | .578301 | .656228 | .706793 | 1.000000 | .873478 | .819428 | .791174 |
| 1 | 3/2 | −.078481 | .091221 | .397437 | .000000 | .309969 | .372379 | .400208 |
| 3 | 5/2 | −.126199 | −.225710 | −.432248 | .000000 | −.363056 | −.403313 | −.413142 |
| 3 | 7/2 | .488574 | .464282 | .288872 | .000000 | −.071797 | −.129133 | −.166357 |
| 5 | 9/2 | −.578164 | −.517510 | −.264894 | .000000 | .062789 | .100511 | .119810 |
| 5 | 11/2 | .265526 | .163472 | .045088 | .000000 | .006983 | .020783 | .034096 |

N = 5 $\mu_B = 0.63$ $\kappa = 0.05$

K = 11/2 - [505]

| | | −0.3 | −0.2 | −0.1 | 0.0 | 0.1 | 0.2 | 0.3 |
|---|---|---|---|---|---|---|---|---|
| E | | 4.831913 | 4.989609 | 5.147304 | 5.305000 | 5.462696 | 5.620392 | 5.778087 |
| L | J | | | | | | | |
| 5 | 11/2 | 1.000000 | 1.000000 | 1.000000 | 1.000000 | 1.000000 | 1.000000 | 1.000000 |

K = 9/2 - [514]

| | | −0.3 | −0.2 | −0.1 | 0.0 | 0.1 | 0.2 | 0.3 |
|---|---|---|---|---|---|---|---|---|
| E | | 5.070230 | 5.154268 | 5.231760 | 5.305000 | 5.375503 | 5.444186 | 5.511618 |
| L | J | | | | | | | |
| 5 | 9/2 | .235020 | .133933 | .057273 | .000000 | −.043232 | −.076504 | −.102661 |
| 5 | 11/2 | .971991 | .990990 | .998359 | 1.000000 | .999065 | .997069 | .994716 |

K = 9/2 - [505]

| | | −0.3 | −0.2 | −0.1 | 0.0 | 0.1 | 0.2 | 0.3 |
|---|---|---|---|---|---|---|---|---|
| E | | 5.427448 | 5.564164 | 5.707466 | 5.855000 | 6.005271 | 6.157363 | 6.310704 |
| L | J | | | | | | | |
| 5 | 9/2 | .971991 | .990990 | .998359 | 1.000000 | .999065 | .997069 | .994716 |
| 5 | 11/2 | −.235020 | −.133933 | −.057273 | .000000 | .043232 | .076504 | .102661 |

| | $\beta =$ | -0.3 | -0.2 | -0.1 | 0.0 | 0.1 | 0.2 | 0.3 |
|---|---|---|---|---|---|---|---|---|
| K = 7/2-[523] | | | | | | | | |
| E | | 5.215928 | 5.268329 | 5.295458 | 5.305000 | 5.302772 | 5.292658 | 5.277188 |
| L | J | | | | | | | |
| 3 | 7/2 | -.403379 | -.231732 | -.096673 | .000000 | .068080 | .116650 | .152100 |
| 5 | 9/2 | .095522 | .088737 | .050531 | .000000 | -.049855 | -.093760 | -.130681 |
| 5 | 11/2 | .910034 | .968724 | .994033 | 1.000000 | .996433 | .988738 | .979688 |
| K = 7/2-[514] | | | | | | | | |
| E | | 5.563549 | 5.677593 | 5.790853 | 5.855000 | 5.903576 | 5.953915 | 6.006885 |
| L | J | | | | | | | |
| 3 | 7/2 | .767266 | .784362 | .492012 | .000000 | -.107764 | -.131268 | -.133505 |
| 5 | 9/2 | .577215 | .606072 | .870561 | 1.000000 | .992540 | .985345 | .979420 |
| 5 | 11/2 | .279509 | .132112 | .003594 | .000000 | .057023 | .108925 | .151373 |
| K = 7/2-[503] | | | | | | | | |
| E | | 5.784819 | 5.807608 | 5.856453 | 5.972000 | 6.114888 | 6.263897 | 6.415632 |
| L | J | | | | | | | |
| 3 | 7/2 | -.498586 | -.575393 | .865204 | 1.000000 | .991843 | .984460 | .979307 |
| 5 | 9/2 | .810986 | .790445 | -.489423 | .000000 | .111262 | .142496 | .153817 |
| 5 | 11/2 | -.306127 | -.210048 | .109023 | .000000 | -.062200 | -.102633 | -.131524 |
| K = 5/2-[532] | | | | | | | | |
| E | | 5.333976 | 5.352249 | 5.341836 | 5.305000 | 5.245981 | 5.169729 | 5.080751 |
| L | J | | | | | | | |
| 3 | 5/2 | -.120050 | -.047602 | -.010220 | .000000 | -.006902 | -.022185 | -.040066 |
| 3 | 7/2 | -.401707 | -.273373 | -.132563 | .000000 | .111308 | .198301 | .263884 |
| 5 | 9/2 | .097002 | .068628 | .038417 | .000000 | -.045112 | -.091331 | -.133994 |
| 5 | 11/2 | .902668 | .958275 | .990377 | 1.000000 | .992737 | .975625 | .954361 |
| K = 5/2-[523] | | | | | | | | |
| E | | 5.763739 | 5.840256 | 5.864961 | 5.855000 | 5.820737 | 5.778277 | 5.735503 |
| L | J | | | | | | | |
| 3 | 5/2 | -.680290 | -.422139 | -.156966 | .000000 | .113327 | .181100 | .221239 |
| 3 | 7/2 | -.020576 | .119114 | .192095 | .000000 | -.140072 | -.179657 | -.177528 |
| 5 | 9/2 | .711186 | .897210 | .968648 | 1.000000 | .981735 | .958093 | .939846 |
| 5 | 11/2 | -.176057 | -.051244 | -.013481 | .000000 | .061105 | .130324 | .190331 |
| K = 5/2-[512] | | | | | | | | |
| E | | 5.911692 | 5.928609 | 5.956782 | 5.972000 | 5.999916 | 6.040539 | 6.088433 |
| L | J | | | | | | | |
| 3 | 5/2 | .357801 | .431476 | .223107 | .000000 | -.083683 | -.120081 | -.143324 |
| 3 | 7/2 | .749472 | .857954 | .953404 | 1.000000 | .978816 | .951007 | .928873 |
| 5 | 9/2 | .446412 | .103886 | -.151029 | .000000 | .155742 | .224827 | .255170 |
| 5 | 11/2 | .333145 | .258748 | .135775 | .000000 | -.103252 | -.174981 | -.227027 |
| K = 5/2-[503] | | | | | | | | |
| E | | 6.255357 | 6.206729 | 6.227343 | 6.322000 | 6.450445 | 6.591612 | 6.738548 |

| β = | | −0.3 | −0.2 | −0.1 | 0.0 | 0.1 | 0.2 | 0.3 |
|---|---|---|---|---|---|---|---|---|
| L | J | | | | | | | |
| 3 | 5/2 | .628309 | .795840 | .962019 | 1.000000 | .990003 | .975854 | .963798 |
| 3 | 7/2 | −.525832 | −.418322 | −.191175 | .000000 | .099548 | .154872 | .189852 |
| 5 | 9/2 | .534341 | .423690 | .193482 | .000000 | −.099530 | −.152214 | −.183366 |
| 5 | 11/2 | −.207866 | −.110147 | −.023167 | .000000 | −.008801 | −.023538 | −.037777 |

K = 3/2-[541]

| | | | | | | | | |
|---|---|---|---|---|---|---|---|---|
| E | | 5.404707 | 5.406053 | 5.372163 | 5.305000 | 5.206737 | 5.080676 | 4.931332 |
| L | J | | | | | | | |
| 1 | 3/2 | .173940 | .072514 | .016373 | .000000 | .012112 | .039964 | .072897 |
| 3 | 5/2 | −.025583 | −.020996 | −.007036 | .000000 | −.008334 | −.031951 | −.065149 |
| 3 | 7/2 | −.432322 | −.302012 | −.153338 | .000000 | .143765 | .266191 | .361868 |
| 5 | 9/2 | .039063 | .038349 | .023594 | .000000 | −.031612 | −.070131 | −.111509 |
| 5 | 11/2 | .883551 | .949536 | .987731 | 1.000000 | .988998 | .960003 | .920358 |

K = 3/2-[532]

| | | | | | | | | |
|---|---|---|---|---|---|---|---|---|
| E | | 5.831517 | 5.902521 | 5.907217 | 5.855000 | 5.760980 | 5.638431 | 5.504591 |
| L | J | | | | | | | |
| 1 | 3/2 | .560749 | −.331333 | −.065841 | .000000 | −.036988 | −.090077 | −.119927 |
| 3 | 5/2 | .443781 | −.378372 | −.208064 | .000000 | .184682 | .313055 | .388614 |
| 3 | 7/2 | −.283216 | .278238 | .123713 | .000000 | −.138483 | −.202657 | −.193943 |
| 5 | 9/2 | −.603777 | .815094 | .968014 | 1.000000 | .970835 | .913248 | .865676 |
| 5 | 11/2 | −.209427 | .072514 | −.004308 | .000000 | .053171 | .137077 | .218146 |

K = 3/2-[521]

| | | | | | | | | |
|---|---|---|---|---|---|---|---|---|
| E | | 6.001957 | 6.007323 | 6.012136 | 5.972000 | 5.907749 | 5.846844 | 5.794302 |
| L | J | | | | | | | |
| 1 | 3/2 | .524623 | −.543237 | −.307784 | .000000 | .184034 | .275250 | .324397 |
| 3 | 5/2 | −.287144 | .152095 | .077845 | .000000 | −.080943 | −.105622 | −.114441 |
| 3 | 7/2 | −.418621 | .667240 | .927903 | 1.000000 | .955673 | .877877 | .811676 |
| 5 | 9/2 | .591372 | −.402303 | −.122110 | .000000 | .166208 | .293727 | .356220 |
| 5 | 11/2 | −.342571 | .273321 | .152623 | .000000 | −.136544 | −.236935 | −.309772 |

K = 3/2-[512]

| | | | | | | | | |
|---|---|---|---|---|---|---|---|---|
| E | | 6.738763 | 6.566915 | 6.427556 | 6.322000 | 6.302841 | 6.315247 | 6.347055 |
| L | J | | | | | | | |
| 1 | 3/2 | −.455436 | −.499844 | −.596135 | .000000 | −.268363 | −.260412 | −.234553 |
| 3 | 5/2 | .674966 | .726781 | .751715 | 1.000000 | .935837 | .889069 | .854333 |
| 3 | 7/2 | −.457729 | −.378212 | −.236774 | .000000 | .157199 | .263063 | .322393 |
| 5 | 9/2 | .332745 | .270144 | .151174 | .000000 | −.164898 | −.264298 | −.321937 |
| 5 | 11/2 | −.129475 | −.076963 | −.025132 | .000000 | −.016949 | −.051820 | −.086712 |

K = 3/2-[501]

| | | | | | | | | |
|---|---|---|---|---|---|---|---|---|
| E | | 6.337145 | 6.273579 | 6.279623 | 6.387000 | 6.504997 | 6.644411 | 6.790633 |
| L | J | | | | | | | |
| 1 | 3/2 | .415516 | .583104 | .738436 | 1.000000 | .944774 | .920170 | .905569 |
| 3 | 5/2 | .514172 | .552313 | .620906 | .000000 | .288929 | .315238 | .318988 |
| 3 | 7/2 | .590023 | .493071 | .210040 | .000000 | −.148769 | −.219550 | −.262073 |
| 5 | 9/2 | .416518 | .315159 | .156933 | .000000 | −.040801 | −.070215 | −.087372 |
| 5 | 11/2 | .203371 | .111782 | .021040 | .000000 | .011186 | .027933 | .043310 |

K = 1/2-[550]

| β = | −0.3 | −0.2 | −0.1 | 0.0 | 0.1 | 0.2 | 0.3 |
|---|---|---|---|---|---|---|---|
| E | 5.441717 | 5.432812 | 5.387174 | 5.305000 | 5.186587 | 5.032459 | 4.843884 |
| a | −5.457371 | −5.771561 | −5.944356 | −6.000000 | −5.944421 | −5.772493 | −5.469689 |
| L J | | | | | | | |
| 1 1/2 | .065258 | .019692 | .002425 | .000000 | −.002133 | −.015164 | −.043585 |
| 1 3/2 | .168616 | .080652 | .020919 | .000000 | .020220 | .075207 | .151284 |
| 3 5/2 | −.032579 | −.011586 | −.002761 | .000000 | −.004086 | −.020488 | −.053804 |
| 3 7/2 | −.437773 | −.313263 | −.162914 | .000000 | .161724 | .308553 | .428177 |
| 5 9/2 | .023838 | .014240 | .007989 | .000000 | −.011488 | −.028739 | −.054557 |
| 5 11/2 | .879793 | .945852 | .986379 | 1.000000 | .986551 | .947451 | .886570 |

K = 1/2-[541]

| β = | −0.3 | −0.2 | −0.1 | 0.0 | 0.1 | 0.2 | 0.3 |
|---|---|---|---|---|---|---|---|
| E | 5.894764 | 5.945828 | 5.928398 | 5.855000 | 5.729705 | 5.551217 | 5.331398 |
| a | 2.861016 | 4.355028 | 4.877027 | 5.000000 | 4.821558 | 3.995302 | 3.035071 |
| L J | | | | | | | |
| 1 1/2 | .443923 | .219307 | .048020 | .000000 | .042220 | .145790 | .236507 |
| 1 3/2 | .218600 | .067990 | −.008172 | .000000 | −.037040 | −.155684 | −.243434 |
| 3 5/2 | −.531509 | −.434301 | −.230142 | .000000 | .227207 | .410387 | .509620 |
| 3 7/2 | −.119088 | −.048272 | .028270 | .000000 | −.075523 | −.178123 | −.179962 |
| 5 9/2 | .655017 | .868523 | .971519 | 1.000000 | .968953 | .861835 | .738813 |
| 5 11/2 | −.171509 | −.044747 | −.003788 | .000000 | .025455 | .107716 | .216473 |

K = 1/2-[530]

| β = | −0.3 | −0.2 | −0.1 | 0.0 | 0.1 | 0.2 | 0.3 |
|---|---|---|---|---|---|---|---|
| E | 6.046113 | 6.051326 | 6.040748 | 5.972000 | 5.848835 | 5.694661 | 5.542161 |
| a | −1.111670 | −2.890448 | −3.794308 | −4.000000 | −3.746713 | −2.743538 | −1.746713 |
| L J | | | | | | | |
| 1 1/2 | −.295023 | −.276763 | −.086548 | .000000 | −.059026 | −.136257 | −.165516 |
| 1 3/2 | −.542817 | −.523238 | −.319123 | .000000 | .299020 | .462712 | .512880 |
| 3 5/2 | −.111934 | .049802 | .045319 | .000000 | −.045329 | −.028430 | .035012 |
| 3 7/2 | .501436 | .726199 | .928773 | 1.000000 | .932600 | .767837 | .611115 |
| 5 9/2 | .475282 | .190979 | −.014071 | .000000 | .101476 | .313872 | .452406 |
| 5 11/2 | .358401 | .288627 | .160621 | .000000 | −.158142 | −.280062 | −.360833 |

K = 1/2-[521]

| β = | −0.3 | −0.2 | −0.1 | 0.0 | 0.1 | 0.2 | 0.3 |
|---|---|---|---|---|---|---|---|
| E | 6.385263 | 6.331265 | 6.343416 | 6.322000 | 6.181479 | 6.054562 | 5.968565 |
| a | .503357 | 1.006475 | 1.827231 | 3.000000 | 1.914814 | 1.403492 | 1.123370 |
| L J | | | | | | | |
| 1 1/2 | .606327 | .680961 | −.618467 | .000000 | .406065 | .480183 | .483978 |
| 1 3/2 | −.080731 | .094201 | −.239563 | .000000 | −.339394 | −.255866 | −.151247 |
| 3 5/2 | −.305323 | −.450968 | .701292 | 1.000000 | .800925 | .657157 | .565678 |
| 3 7/2 | .500566 | .411785 | −.167557 | .000000 | .191692 | .374504 | .453499 |
| 5 9/2 | −.482769 | −.376049 | .199478 | .000000 | −.202938 | −.351669 | −.437033 |
| 5 11/2 | .221348 | .114309 | −.020727 | .000000 | −.022636 | −.090424 | −.161984 |

K = 1/2-[510]

| β = | −0.3 | −0.2 | −0.1 | 0.0 | 0.1 | 0.2 | 0.3 |
|---|---|---|---|---|---|---|---|
| E | 6.780222 | 6.602287 | 6.458236 | 6.387000 | 6.336110 | 6.343453 | 6.373591 |
| a | −.014041 | −.049350 | −.498715 | −2.000000 | −.862397 | −.601867 | −.590911 |
| L J | | | | | | | |
| 1 1/2 | .180359 | .188617 | .258519 | .000000 | −.169667 | −.164678 | −.171147 |
| 1 3/2 | .546286 | .626107 | .763685 | 1.000000 | .826598 | .731608 | .675725 |

| L | J | $\beta$ = −0.3 | −0.2 | −0.1 | 0.0 | 0.1 | 0.2 | 0.3 |
|---|---|---|---|---|---|---|---|---|
| 3 | 5/2 | .594742 | .600779 | .520198 | .000000 | .471182 | .525581 | .526599 |
| 3 | 7/2 | .461203 | .391196 | .258202 | .000000 | -.239133 | -.357762 | -.423518 |
| 5 | 9/2 | .294751 | .229443 | .109465 | .000000 | -.090733 | -.171939 | -.218078 |
| 5 | 11/2 | .125450 | .076153 | .026384 | .000000 | .022787 | .061952 | .099360 |

K = 1/2-[501]

| L | J | −0.3 | −0.2 | −0.1 | 0.0 | 0.1 | 0.2 | 0.3 |
|---|---|---|---|---|---|---|---|---|
| E | | 7.249183 | 6.960657 | 6.693116 | 6.537000 | 6.622197 | 6.755474 | 6.899140 |
| a | | .218710 | .349855 | .533120 | 1.000000 | .817160 | .719104 | .648872 |
| 1 | 1/2 | -.558098 | -.612891 | .735434 | 1.000000 | .895016 | .838006 | .806996 |
| 1 | 3/2 | .569374 | .560545 | -.507002 | .000000 | .332194 | .394063 | .418721 |
| 3 | 5/2 | -.506919 | -.494429 | .427265 | .000000 | -.287772 | -.349662 | -.372652 |
| 3 | 7/2 | .281886 | .222642 | -.123680 | .000000 | -.066849 | -.123479 | -.160588 |
| 5 | 9/2 | -.156676 | -.122208 | .064155 | .000000 | .035830 | .068300 | .089169 |
| 5 | 11/2 | .058010 | .034485 | -.010807 | .000000 | .005311 | .016856 | .028652 |

N = 5 $\mu_5$ = 0.7 $\kappa$ = 0.05 odd proton nuclei

K = 11/2-[505]

| L | J | −0.3 | −0.2 | −0.1 | 0.0 | 0.1 | 0.2 | 0.3 |
|---|---|---|---|---|---|---|---|---|
| E | | 4.726913 | 4.884609 | 5.042304 | 5.200000 | 5.357696 | 5.515392 | 5.673087 |
| 5 | 11/2 | 1.000000 | 1.000000 | 1.000000 | 1.000000 | 1.000000 | 1.000000 | 1.000000 |

K = 9/2-[514]

| L | J | −0.3 | −0.2 | −0.1 | 0.0 | 0.1 | 0.2 | 0.3 |
|---|---|---|---|---|---|---|---|---|
| E | | 4.965230 | 5.049288 | 5.126760 | 5.200000 | 5.270503 | 5.339186 | 5.406618 |
| 5 | 9/2 | .235020 | .133933 | .057273 | .000000 | -.043232 | -.076504 | -.102661 |
| 5 | 11/2 | .971991 | .990990 | .998359 | 1.000000 | .999065 | .997069 | .994716 |

K = 9/2-[505]

| L | J | −0.3 | −0.2 | −0.1 | 0.0 | 0.1 | 0.2 | 0.3 |
|---|---|---|---|---|---|---|---|---|
| E | | 5.322448 | 5.459164 | 5.602466 | 5.750000 | 5.900271 | 6.052363 | 6.205705 |
| 5 | 9/2 | .971991 | .990990 | .998359 | 1.000000 | .999065 | .997069 | .994716 |
| 5 | 11/2 | -.235020 | -.133933 | -.057273 | .000000 | .043232 | .076504 | .102661 |

K = 7/2-[523]

| L | J | −0.3 | −0.2 | −0.1 | 0.0 | 0.1 | 0.2 | 0.3 |
|---|---|---|---|---|---|---|---|---|
| E | | 5.119940 | 5.166312 | 5.190986 | 5.200000 | 5.198043 | 5.188464 | 5.173570 |
| 3 | 7/2 | -.354094 | -.204214 | -.080701 | .000000 | .063179 | .109573 | .144210 |
| 5 | 9/2 | .108447 | .092844 | .051130 | .000000 | -.049686 | -.093403 | -.130249 |
| 5 | 11/2 | .928901 | .974514 | .994921 | 1.000000 | .996765 | .989581 | .980938 |

K = 7/2-[514]

| L | J | −0.3 | −0.2 | −0.1 | 0.0 | 0.1 | 0.2 | 0.3 |
|---|---|---|---|---|---|---|---|---|
| E | | 5.493626 | 5.604746 | 5.694220 | 5.750000 | 5.799141 | 5.849823 | 5.902864 |
| 3 | 7/2 | .719467 | .632416 | .261416 | .000000 | -.083281 | -.109706 | -.116461 |

| | β = | −0.3 | −0.2 | −0.1 | 0.0 | 0.1 | 0.2 | 0.3 |
|---|---|---|---|---|---|---|---|---|
| 5 | 9/2 | .666154 | .772383 | .964854 | 1.000000 | .995014 | .988356 | .982176 |
| 5 | 11/2 | .196487 | .058939 | −.026804 | .000000 | .054878 | .105435 | .147535 |

K = 7/2−[503]

| | | | | | | | | |
|---|---|---|---|---|---|---|---|---|
| E | | 5.698730 | 5.730472 | 5.805559 | 5.930000 | 6.072051 | 6.220184 | 6.371270 |
| L | J | | | | | | | |
| 3 | 7/2 | −.597482 | .747226 | .961324 | 1.000000 | .994521 | .987906 | .982670 |
| 5 | 9/2 | .737888 | −.628334 | −.257764 | .000000 | .086478 | .120116 | .135517 |
| 5 | 11/2 | −.313905 | .216448 | .097020 | .000000 | −.058726 | −.098050 | −.126471 |

K = 5/2−[532]

| | | | | | | | | |
|---|---|---|---|---|---|---|---|---|
| E | | 5.239149 | 5.251663 | 5.237847 | 5.200000 | 5.141704 | 5.067071 | 4.979986 |
| L | J | | | | | | | |
| 3 | 5/2 | −.104970 | −.041449 | −.008944 | .000000 | −.006218 | −.020315 | −.037227 |
| 3 | 7/2 | −.370280 | −.249005 | −.120425 | .000000 | .102772 | .185212 | .249064 |
| 5 | 9/2 | .099350 | .070440 | .038944 | .000000 | −.044780 | −.090496 | −.132915 |
| 5 | 11/2 | .917607 | .965047 | .991918 | 1.000000 | .993677 | .978312 | .958600 |

K = 5/2−[523]

| | | | | | | | | |
|---|---|---|---|---|---|---|---|---|
| E | | 5.684878 | 5.745737 | 5.762916 | 5.750000 | 5.717369 | 5.676816 | 5.635084 |
| L | J | | | | | | | |
| 3 | 5/2 | −.606519 | −.357465 | −.143276 | .000000 | .101081 | .165829 | .206527 |
| 3 | 7/2 | .024905 | .126297 | .126524 | .000000 | −.103038 | −.145253 | −.151999 |
| 5 | 9/2 | .781532 | .923984 | .981257 | 1.000000 | .987953 | .967929 | .949804 |
| 5 | 11/2 | −.143950 | −.050208 | −.024457 | .000000 | .055811 | .120478 | .179209 |

K = 5/2−[512]

| | | | | | | | | |
|---|---|---|---|---|---|---|---|---|
| E | | 5.852368 | 5.882307 | 5.912862 | 5.930000 | 5.956132 | 5.994045 | 6.039696 |
| L | J | | | | | | | |
| 3 | 5/2 | .445433 | .448549 | .208330 | .000000 | −.089564 | −.129953 | −.153710 |
| 3 | 7/2 | .748601 | .859292 | .966042 | 1.000000 | .984274 | .959295 | .937159 |
| 5 | 9/2 | .379290 | .068898 | −.091085 | .000000 | .117302 | .187010 | .224592 |
| 5 | 11/2 | .311971 | .235954 | .122738 | .000000 | −.097074 | −.167012 | −.218321 |

K = 5/2−[503]

| | | | | | | | | |
|---|---|---|---|---|---|---|---|---|
| E | | 6.194371 | 6.154136 | 6.183297 | 6.280000 | 6.407873 | 6.548224 | 6.694469 |
| L | J | | | | | | | |
| 3 | 5/2 | .650158 | .818110 | .967466 | 1.000000 | .990819 | .977343 | .965574 |
| 3 | 7/2 | −.549426 | −.428559 | −.190399 | .000000 | .100129 | .156049 | .191300 |
| 5 | 9/2 | .485259 | .369519 | .165292 | .000000 | −.090466 | −.141247 | −.172525 |
| 5 | 11/2 | −.199873 | −.102412 | −.020881 | .000000 | −.008233 | −.022313 | −.036128 |

K = 3/2−[541]

| | | | | | | | | |
|---|---|---|---|---|---|---|---|---|
| E | | 5.313139 | 5.306781 | 5.268541 | 5.200000 | 5.102950 | 4.980015 | 4.834793 |
| L | J | | | | | | | |
| 1 | 3/2 | .143664 | .059698 | .013556 | .000000 | .010326 | .034795 | .064859 |
| 3 | 5/2 | −.028219 | −.019986 | −.006353 | .000000 | −.007383 | −.028603 | −.059254 |
| 3 | 7/2 | −.401033 | −.277191 | −.140050 | .000000 | .132131 | .247144 | .340022 |

| β = | | −0.3 | −0.2 | −0.1 | 0.0 | 0.1 | 0.2 | 0.3 |
|---|---|---|---|---|---|---|---|---|
| 5 | 9/2 | .044658 | .040193 | .023981 | .000000 | −.031234 | −.068869 | −.109538 |
| 5 | 11/2 | .903185 | .957907 | .989741 | 1.000000 | .990659 | .965478 | .929876 |

**K = 3/2 - [532]**

| | | | | | | | | |
|---|---|---|---|---|---|---|---|---|
| E | | 5.769043 | 5.816238 | 5.805537 | 5.750000 | 5.658805 | 5.541667 | 5.411670 |
| L | J | | | | | | | |
| 1 | 3/2 | −.450568 | −.207714 | −.041024 | .000000 | −.025301 | −.068737 | −.099562 |
| 3 | 5/2 | −.452861 | −.357296 | −.185148 | .000000 | .163649 | .285378 | .363466 |
| 3 | 7/2 | .247891 | .178740 | .079839 | .000000 | −.094595 | −.155317 | −.162973 |
| 5 | 9/2 | .716229 | .892658 | .978516 | 1.000000 | .980614 | .935904 | .890807 |
| 5 | 11/2 | .132175 | .019757 | −.013038 | .000000 | .045018 | .117450 | .194635 |

**K = 3/2 - [521]**

| | | | | | | | | |
|---|---|---|---|---|---|---|---|---|
| E | | 5.945032 | 5.965126 | 5.971270 | 5.930000 | 5.864539 | 5.799676 | 5.743116 |
| L | J | | | | | | | |
| 1 | 3/2 | .602852 | −.562121 | −.282169 | .000000 | .175569 | .269785 | .321087 |
| 3 | 5/2 | −.189375 | .081322 | .071042 | .000000 | −.091586 | −.130687 | −.142651 |
| 3 | 7/2 | −.489924 | .738760 | .943627 | 1.000000 | .964602 | .896087 | .830916 |
| 5 | 9/2 | .492498 | −.251951 | −.073520 | .000000 | .118715 | .236746 | .311358 |
| 5 | 11/2 | −.343697 | .261076 | .139627 | .000000 | −.127426 | −.226088 | −.298644 |

**K = 3/2 - [512]**

| | | | | | | | | |
|---|---|---|---|---|---|---|---|---|
| E | | 6.288690 | 6.237173 | 6.253609 | 6.280000 | 6.261445 | 6.271183 | 6.300264 |
| L | J | | | | | | | |
| 1 | 3/2 | .422286 | .585494 | .671805 | .000000 | −.231190 | −.237333 | −.218485 |
| 3 | 5/2 | .560808 | .606840 | .703117 | 1.000000 | .949685 | .903439 | .867856 |
| 3 | 7/2 | .566951 | .444231 | .157159 | .000000 | .148650 | .255796 | .317627 |
| 5 | 9/2 | .390286 | .288130 | .149513 | .000000 | −.149422 | −.244485 | −.302657 |
| 5 | 11/2 | .182792 | .092631 | .013960 | .000000 | −.015051 | −.047600 | −.081256 |

**K = 3/2 - [501]**

| | | | | | | | | |
|---|---|---|---|---|---|---|---|---|
| E | | 6.697183 | 6.530074 | 6.398740 | 6.380000 | 6.494565 | 6.632068 | 6.777069 |
| L | J | | | | | | | |
| 1 | 3/2 | −.484348 | −.542686 | −.683514 | 1.000000 | .956546 | .930029 | .913810 |
| 3 | 5/2 | .660147 | .705033 | .677645 | .000000 | .250750 | .290620 | .301427 |
| 3 | 7/2 | −.465042 | −.384849 | −.242652 | .000000 | −.145049 | −.215388 | −.257909 |
| 5 | 9/2 | .300240 | .234585 | .119048 | .000000 | −.031629 | −.059318 | −.076935 |
| 5 | 11/2 | −.123478 | −.072676 | −.023509 | .000000 | .010247 | .025997 | .040714 |

**K = 1/2 - [550]**

| | | | | | | | | |
|---|---|---|---|---|---|---|---|---|
| E | | 5.350789 | 5.334092 | 5.283742 | 5.200000 | 5.083132 | 4.933514 | 4.751960 |
| a | | −5.550399 | −5.809603 | −5.953465 | −6.000000 | −5.953387 | −5.808908 | −5.552027 |
| L | J | | | | | | | |
| 1 | 1/2 | .051821 | .015440 | .001893 | .000000 | −.001686 | −.012195 | −.035967 |
| 1 | 3/2 | .144838 | .068240 | .017554 | .000000 | .017026 | .064094 | .131409 |
| 3 | 5/2 | −.027250 | −.010061 | −.002465 | .000000 | −.003544 | −.017553 | −.046358 |
| 3 | 7/2 | −.408907 | −.288872 | −.149162 | .000000 | .148182 | .284967 | .400620 |
| 5 | 9/2 | .022957 | .014448 | .008114 | .000000 | −.011298 | −.027687 | −.051794 |
| 5 | 11/2 | .898811 | .954645 | .983619 | 1.000000 | .988741 | .955752 | .903388 |

| L | J | β = -0.3 | -0.2 | -0.1 | 0.0 | 0.1 | 0.2 | 0.3 |
|---|---|---|---|---|---|---|---|---|

K = 1/2-[541]

| | | | | | | | | |
|---|---|---|---|---|---|---|---|---|
| E | | 5.826960 | 5.855241 | 5.826570 | 5.750000 | 5.628017 | 5.460487 | 5.253027 |
| a | | 3.598804 | 4.579385 | 4.906842 | 5.000000 | 4.888578 | 4.404323 | 3.593494 |
| L | J | | | | | | | |
| 1 | 1/2 | .361061 | .161016 | .036913 | .000000 | .032139 | .114074 | .201661 |
| 1 | 3/2 | .145811 | .027625 | -.006392 | .000000 | -.022913 | -.104078 | -.191911 |
| 3 | 5/2 | -.516407 | -.389913 | -.203705 | .000000 | .200841 | .373226 | .484974 |
| 3 | 7/2 | -.077332 | -.008490 | .021409 | .000000 | -.045806 | -.117713 | -.146216 |
| 5 | 9/2 | .750101 | .905861 | .978067 | 1.000000 | .977567 | .903957 | .797835 |
| 5 | 11/2 | -.114310 | -.024967 | -.005262 | .000000 | .019204 | .076573 | .171415 |

K = 1/2-[530]

| | | | | | | | | |
|---|---|---|---|---|---|---|---|---|
| E | | 5.999068 | 6.014952 | 6.000794 | 5.930000 | 5.808012 | 5.651512 | 5.491620 |
| a | | -1.677910 | -3.086712 | -3.825005 | -4.000000 | -3.810314 | -3.110220 | -2.190430 |
| L | J | | | | | | | |
| 1 | 1/2 | .356538 | -.272494 | -.075589 | .000000 | -.054765 | -.143224 | -.185960 |
| 1 | 3/2 | .557579 | -.505612 | -.296907 | .000000 | .280784 | .456683 | .521649 |
| 3 | 5/2 | .011915 | .088013 | .042490 | .000000 | -.053222 | -.075889 | -.029526 |
| 3 | 7/2 | -.543910 | .759110 | .939424 | 1.000000 | .943233 | .801232 | .649198 |
| 5 | 9/2 | -.380875 | .116278 | -.010008 | .000000 | .066376 | .228866 | .384739 |
| 5 | 11/2 | -.347765 | .269421 | .147344 | .000000 | -.145722 | -.266113 | -.350637 |

K = 1/2-[521]

| | | | | | | | | |
|---|---|---|---|---|---|---|---|---|
| E | | 6.340018 | 6.296978 | 6.312832 | 6.280000 | 6.145816 | 6.014906 | 5.922567 |
| a | | .512747 | 1.157718 | 2.151970 | 3.000000 | 2.191017 | 1.567235 | 1.208173 |
| L | J | | | | | | | |
| 1 | 1/2 | .611584 | .674828 | -.562676 | .000000 | .374816 | .464823 | .475197 |
| 1 | 3/2 | -.075562 | .096351 | -.177923 | .000000 | -.284340 | -.239260 | -.144914 |
| 3 | 5/2 | -.362247 | -.514479 | .774657 | 1.000000 | .846828 | .701473 | .605704 |
| 3 | 7/2 | .497939 | .383127 | -.132185 | .000000 | .164391 | .351867 | .442761 |
| 5 | 9/2 | -.446682 | -.338050 | .184224 | .000000 | -.184908 | -.323282 | -.410045 |
| 5 | 11/2 | .203875 | .097825 | -.015288 | .000000 | -.018180 | -.079419 | -.148777 |

K = 1/2-[510]

| | | | | | | | | |
|---|---|---|---|---|---|---|---|---|
| E | | 6.744149 | 6.572466 | 6.439727 | 6.380000 | 6.319674 | 6.319946 | 6.345771 |
| a | | -.009569 | -.079981 | -.703423 | -2.000000 | -1.096607 | -.731391 | -.672064 |
| L | J | | | | | | | |
| 1 | 1/2 | .213449 | .238608 | .343695 | .000000 | -.213196 | -.198486 | -.196788 |
| 1 | 3/2 | .548907 | .630380 | .779188 | 1.000000 | .851588 | .747576 | .687450 |
| 3 | 5/2 | .601745 | .596239 | .452569 | .000000 | .411242 | .502255 | .515901 |
| 3 | 7/2 | .452717 | .380976 | .250583 | .000000 | -.234610 | -.352894 | -.419122 |
| 5 | 9/2 | .269264 | .200494 | .081021 | .000000 | -.068923 | -.147077 | -.195354 |
| 5 | 11/2 | .116567 | .069611 | .023777 | .000000 | .020820 | .057517 | .093306 |

K = 1/2-[501]

| | | | | | | | | |
|---|---|---|---|---|---|---|---|---|
| E | | 7.228277 | 6.942446 | 6.679424 | 6.530000 | 6.612262 | 6.743460 | 6.885794 |
| a | | .126327 | .239194 | .423081 | 1.000000 | .780714 | .678961 | .612853 |
| L | J | | | | | | | |
| 1 | 1/2 | -.565893 | -.622298 | .747121 | 1.000000 | .900014 | .843130 | .811739 |
| 1 | 3/2 | .582974 | .576431 | -.522213 | .000000 | .338075 | .400482 | .424494 |
| 3 | 5/2 | -.488998 | -.468969 | .389591 | .000000 | -.265668 | -.332130 | -.358815 |

| β = | -0.3 | -0.2 | -0.1 | 0.0 | 0.1 | 0.2 | 0.3 |
|---|---|---|---|---|---|---|---|
| 3 7/2 | .279429 | .219780 | -.120461 | .000000 | -.064728 | -.120909 | -.158003 |
| 5 9/2 | -.140458 | -.105883 | .052116 | .000000 | .029789 | .059777 | .080322 |
| 5 11/2 | .054570 | .032025 | -.009789 | .000000 | .004802 | .015583 | .026831 |

**N = 6 $\mu_6$ = 0.4476 $\kappa$ = 0.05 odd neutron nuclei**

**K = 13/2+[606]**

| | | | | | | | |
|---|---|---|---|---|---|---|---|
| E | 5.692335 | 5.881570 | 6.070805 | 6.260040 | 6.449275 | 6.638510 | 6.827745 |
| L J | | | | | | | |
| 6 13/2 | 1.000000 | 1.000000 | 1.000000 | 1.000000 | 1.000000 | 1.000000 | 1.000000 |

**K = 11/2+[615]**

| | | | | | | | |
|---|---|---|---|---|---|---|---|
| E | 5.940838 | 6.051112 | 6.157031 | 6.260040 | 6.361066 | 6.460705 | 6.559346 |
| L J | | | | | | | |
| 6 11/2 | .175888 | .101342 | .044108 | .000000 | -.034473 | -.061886 | -.084066 |
| 6 13/2 | .984410 | .994852 | .999027 | 1.000000 | .999406 | .998083 | .996460 |

**K = 11/2+[606]**

| | | | | | | | |
|---|---|---|---|---|---|---|---|
| E | 6.377665 | 6.551263 | 6.729197 | 6.910040 | 7.092866 | 7.277080 | 7.462291 |
| L J | | | | | | | |
| 6 11/2 | .984410 | .994852 | .999027 | 1.000000 | .999406 | .998083 | .996460 |
| 6 13/2 | -.175888 | -.101342 | -.044108 | .000000 | .034473 | .061886 | .084066 |

**K = 9/2+[624]**

| | | | | | | | |
|---|---|---|---|---|---|---|---|
| E | 6.080739 | 6.169010 | 6.224348 | 6.260040 | 6.284549 | 6.302295 | 6.315676 |
| L J | | | | | | | |
| 4 9/2 | -.491958 | -.273717 | -.107031 | .000000 | .066611 | .114567 | .146814 |
| 6 11/2 | .070921 | .072781 | .042226 | .000000 | -.041380 | -.077879 | -.108939 |
| 6 13/2 | .867726 | .959052 | .993359 | 1.000000 | .996785 | .990358 | .983147 |

**K = 9/2+[604]**

| | | | | | | | |
|---|---|---|---|---|---|---|---|
| E | 6.390812 | 6.523807 | 6.681864 | 6.852400 | 7.036629 | 7.217217 | 7.401346 |
| L J | | | | | | | |
| 4 9/2 | .831516 | .937242 | .981667 | 1.000000 | .909457 | .962914 | .969418 |
| 6 11/2 | .333620 | .244117 | .162974 | .000000 | .413308 | .253846 | .213451 |
| 6 13/2 | .444161 | .248967 | .098844 | .000000 | -.045442 | -.091430 | -.121112 |

**K = 9/2+[615]**

| | | | | | | | |
|---|---|---|---|---|---|---|---|
| E | 6.699372 | 6.761958 | 6.832415 | 6.910040 | 6.985154 | 7.070673 | 7.157015 |
| L J | | | | | | | |
| 4 9/2 | -.257990 | -.216001 | -.157718 | .000000 | -.410099 | -.244277 | -.196660 |
| 6 11/2 | .940036 | .967011 | .985726 | 1.000000 | .909651 | .964104 | .970861 |
| 6 13/2 | -.223099 | -.135033 | -.058896 | .000000 | .065991 | .104073 | .136945 |

| β = | | −0.3 | −0.2 | −0.1 | 0.0 | 0.1 | 0.2 | 0.3 |
|---|---|---|---|---|---|---|---|---|
| **K = 7/2+[633]** | | | | | | | | |
| E | | 6.216160 | 6.262875 | 6.276455 | 6.260040 | 6.220839 | 6.165943 | 6.100510 |
| L | J | | | | | | | |
| 4 | 7/2 | −.115069 | −.045117 | −.009324 | .000000 | −.005662 | −.017510 | −.030847 |
| 4 | 9/2 | −.462005 | −.314417 | −.148605 | .000000 | .114602 | .197435 | .256612 |
| 6 | 11/2 | .081821 | .061328 | .035391 | .000000 | −.040523 | −.080460 | −.116584 |
| 6 | 13/2 | .875566 | .946227 | .988219 | 1.000000 | .992568 | .976851 | .958961 |
| **K = 7/2+[613]** | | | | | | | | |
| E | | 6.745686 | 6.759637 | 6.798619 | 6.852400 | 6.935799 | 7.009301 | 7.092195 |
| L | J | | | | | | | |
| 4 | 7/2 | −.043468 | .214566 | .081715 | .000000 | .013607 | −.042317 | −.075359 |
| 4 | 9/2 | .760817 | .906003 | .968723 | 1.000000 | .758900 | .901153 | .914368 |
| 6 | 11/2 | .548252 | .211078 | .188093 | .000000 | .648194 | .404678 | .340495 |
| 6 | 13/2 | .344510 | .297602 | .139709 | .000000 | −.061081 | −.149562 | −.205708 |
| **K = 7/2+[624]** | | | | | | | | |
| E | | 6.686166 | 6.824420 | 6.896611 | 6.910040 | 6.888559 | 6.882994 | 6.874225 |
| L | J | | | | | | | |
| 4 | 7/2 | .794992 | −.588065 | −.250220 | .000000 | .122317 | .183735 | .215929 |
| 4 | 9/2 | .256308 | −.019565 | −.155441 | .000000 | −.637341 | −.369476 | −.279640 |
| 6 | 11/2 | −.470944 | .803921 | .953751 | 1.000000 | .753525 | .898132 | .915371 |
| 6 | 13/2 | .283734 | −.086646 | −.059892 | .000000 | .105049 | .151946 | .193060 |
| **K = 7/2+[604]** | | | | | | | | |
| E | | 7.109163 | 7.099478 | 7.163960 | 7.302400 | 7.468918 | 7.645112 | 7.825655 |
| L | J | | | | | | | |
| 4 | 7/2 | .594017 | .778527 | .964689 | 1.000000 | .992382 | .981908 | .973008 |
| 4 | 9/2 | −.376848 | −.282698 | −.123812 | .000000 | .068805 | .111494 | .141010 |
| 6 | 11/2 | .686248 | .552627 | .231792 | .000000 | −.101996 | −.152053 | −.180463 |
| 6 | 13/2 | −.184911 | −.092633 | −.017818 | .000000 | −.006447 | −.017458 | −.028373 |
| **K = 5/2+[642]** | | | | | | | | |
| E | | 6.300340 | 6.329667 | 6.314650 | 6.260040 | 6.171196 | 6.055176 | 5.919226 |
| L | J | | | | | | | |
| 2 | 5/2 | .231526 | .091546 | .019089 | .000000 | .011697 | .035462 | .060602 |
| 4 | 7/2 | −.036263 | −.027212 | −.008408 | .000000 | −.008206 | −.028724 | −.054598 |
| 4 | 9/2 | −.492615 | −.348280 | −.175074 | .000000 | .153152 | .272222 | .358431 |
| 6 | 11/2 | .041654 | .041932 | .026193 | .000000 | −.033911 | −.071637 | −.108722 |
| 6 | 13/2 | .837064 | .931570 | .983986 | 1.000000 | .987517 | .958478 | .923609 |
| **K = 5/2+[633]** | | | | | | | | |
| E | | 6.689131 | 6.804451 | 6.863634 | 6.852400 | 6.805556 | 6.717266 | 6.617181 |
| L | J | | | | | | | |
| 2 | 5/2 | .763911 | −.665744 | −.312158 | .000000 | .135410 | −.141717 | −.136779 |
| 4 | 7/2 | .282621 | −.147966 | .007073 | .000000 | −.165881 | .302901 | .368678 |
| 4 | 9/2 | −.326064 | .616994 | .916328 | 1.000000 | .794236 | −.473523 | −.329265 |
| 6 | 11/2 | −.297974 | .275998 | .189521 | .000000 | −.549819 | .787912 | .819700 |
| 6 | 13/2 | −.376111 | .279349 | .164107 | .000000 | −.145040 | .207697 | .255039 |

K = 5/2+[622]

| L | J | β = −0.3 | −0.2 | −0.1 | 0.0 | 0.1 | 0.2 | 0.3 |
|---|---|---|---|---|---|---|---|---|
| E | | 6.908108 | 6.936100 | 6.945416 | 6.910040 | 6.853988 | 6.825622 | 6.810018 |
| 2 | 5/2 | .233879 | .099002 | .016134 | .000000 | .084601 | .198337 | .258142 |
| 4 | 7/2 | −.559774 | −.484908 | −.270170 | .000000 | .100353 | .034266 | −.013386 |
| 4 | 9/2 | −.258320 | −.291335 | −.177127 | .000000 | .560380 | .789929 | .804455 |
| 6 | 11/2 | .699412 | .801043 | .944364 | 1.000000 | .815633 | .547258 | .458109 |
| 6 | 13/2 | −.275766 | −.168869 | −.059275 | .000000 | −.059068 | −.189761 | −.275993 |

K = 5/2+[602]

| L | J | β = −0.3 | −0.2 | −0.1 | 0.0 | 0.1 | 0.2 | 0.3 |
|---|---|---|---|---|---|---|---|---|
| F | | 7.152622 | 7.106903 | 7.141349 | 7.265720 | 7.434758 | 7.609133 | 7.789051 |
| 2 | 5/2 | .460853 | .662117 | .900957 | 1.000000 | .900105 | .916588 | .916362 |
| 4 | 7/2 | .312345 | .311511 | .311187 | .000000 | .415638 | .345346 | .316499 |
| 4 | 9/2 | .658916 | .570857 | .272874 | .000000 | −.111726 | −.182204 | −.225005 |
| 6 | 11/2 | .438606 | .344284 | .126717 | .000000 | −.067174 | −.083195 | −.090784 |
| 6 | 13/2 | .252011 | .141959 | .030358 | .000000 | .007813 | .021969 | .035216 |

K = 5/2+[613]

| L | J | β = −0.3 | −0.2 | −0.1 | 0.0 | 0.1 | 0.2 | 0.3 |
|---|---|---|---|---|---|---|---|---|
| E | | 7.540399 | 7.413460 | 7.325551 | 7.302400 | 7.325103 | 7.383403 | 7.455124 |
| 2 | 5/2 | −.309436 | −.316551 | −.300348 | .000000 | −.405192 | −.314941 | −.266951 |
| 4 | 7/2 | .712674 | .803238 | .911071 | 1.000000 | .888589 | .887123 | .872207 |
| 4 | 9/2 | −.387444 | −.295408 | −.154458 | .000000 | .138655 | .210914 | .255610 |
| 6 | 11/2 | .477413 | .402325 | .235537 | .000000 | −.163646 | −.260097 | −.313316 |
| 6 | 13/2 | −.135307 | −.073961 | −.020139 | .000000 | −.014939 | −.041104 | −.067003 |

K = 3/2+[651]

| L | J | β = −0.3 | −0.2 | −0.1 | 0.0 | 0.1 | 0.2 | 0.3 |
|---|---|---|---|---|---|---|---|---|
| E | | 6.362369 | 6.373971 | 6.339755 | 6.260040 | 6.136968 | 5.974771 | 5.780088 |
| 2 | 3/2 | .078529 | .022864 | .002620 | .000000 | −.001815 | −.011101 | −.027563 |
| 2 | 5/2 | .216634 | .105390 | .026772 | .000000 | .022448 | .075253 | .136568 |
| 4 | 7/2 | −.045769 | −.020648 | −.005840 | .000000 | −.007533 | −.030865 | −.065942 |
| 4 | 9/2 | −.497155 | −.364860 | −.190948 | .000000 | .181807 | .332733 | .442995 |
| 6 | 11/2 | .034354 | .026153 | .016035 | .000000 | −.022634 | −.051422 | −.083803 |
| 6 | 13/2 | .834545 | .924195 | .981083 | 1.000000 | .982787 | .938033 | .879189 |

K = 3/2+[642]

| L | J | β = −0.3 | −0.2 | −0.1 | 0.0 | 0.1 | 0.2 | 0.3 |
|---|---|---|---|---|---|---|---|---|
| E | | 6.837295 | 6.916358 | 6.910987 | 6.852400 | 6.741326 | 6.579491 | 6.391053 |
| 2 | 3/2 | .616585 | −.276329 | −.054119 | .000000 | −.043341 | −.117659 | .164033 |
| 2 | 5/2 | .366620 | −.553500 | −.330945 | .000000 | .253546 | .305998 | −.287936 |
| 4 | 7/2 | −.452886 | .103831 | .006764 | .000000 | −.150788 | −.364984 | .473165 |
| 4 | 9/2 | −.219837 | .697033 | .914355 | 1.000000 | .874537 | .549017 | −.336492 |
| 6 | 11/2 | .356673 | .013938 | .131223 | .000000 | −.339248 | −.621864 | .671826 |
| 6 | 13/2 | −.323670 | .347059 | .185032 | .000000 | −.176622 | −.266784 | .318942 |

K = 3/2+[631]

| L | J | β = −0.3 | −0.2 | −0.1 | 0.0 | 0.1 | 0.2 | 0.3 |
|---|---|---|---|---|---|---|---|---|
| E | | 6.930141 | 6.961464 | 6.974512 | 6.910040 | 6.793780 | 6.672344 | 6.560874 |

| β = | -0.3 | -0.2 | -0.1 | 0.0 | 0.1 | 0.2 | 0.3 |
|---|---|---|---|---|---|---|---|
| L J | | | | | | | |
| 2 3/2 | -.021818 | .284403 | .079097 | .000000 | .019545 | -.004087 | -.041193 |
| 2 5/2 | .497702 | -.063465 | .031844 | .000000 | .085637 | .285519 | .400683 |
| 4 7/2 | .375253 | -.503064 | -.298695 | .000000 | .201957 | .185568 | .136555 |
| 4 9/2 | -.424299 | .115353 | -.109462 | .000000 | .354040 | .626042 | .636923 |
| 6 11/2 | -.563916 | .805358 | .946484 | 1.000000 | .907817 | .671758 | .557066 |
| 6 13/2 | -.336112 | .011712 | -.039577 | .000000 | -.044959 | -.202088 | -.321116 |

**K = 3/2+ [611]**

| | -0.3 | -0.2 | -0.1 | 0.0 | 0.1 | 0.2 | 0.3 |
|---|---|---|---|---|---|---|---|
| E | 7.255446 | 7.236728 | 7.286978 | 7.265720 | 7.298960 | 7.352466 | 7.422593 |
| L J | | | | | | | |
| 2 3/2 | -.648505 | .761396 | .651001 | .000000 | .015172 | -.039629 | -.074878 |
| 2 5/2 | .107439 | .102391 | .604343 | 1.000000 | .773122 | .788644 | .769061 |
| 4 7/2 | .031887 | -.228074 | -.337728 | .000000 | .594513 | .502373 | .466989 |
| 4 9/2 | -.408964 | .365676 | .273378 | .000000 | -.171283 | -.297385 | -.369782 |
| 6 11/2 | .587682 | -.457433 | -.184650 | .000000 | -.137952 | -.182547 | -.204230 |
| 6 13/2 | -.232937 | .121701 | .035351 | .000000 | .015434 | .048273 | .080071 |

**K = 3/2+ [622]**

| | -0.3 | -0.2 | -0.1 | 0.0 | 0.1 | 0.2 | 0.3 |
|---|---|---|---|---|---|---|---|
| E | 7.575106 | 7.426802 | 7.321727 | 7.302400 | 7.196339 | 7.135131 | 7.098343 |
| L J | | | | | | | |
| 2 3/2 | .066705 | .016337 | -.256860 | .000000 | .304817 | .383238 | .409521 |
| 2 5/2 | .585986 | .687312 | .629761 | .000000 | -.547902 | -.377577 | -.274935 |
| 4 7/2 | .486331 | .498720 | .675409 | 1.000000 | .720897 | .692759 | .647421 |
| 4 9/2 | .511650 | .426125 | .171693 | .000000 | .215263 | .320397 | .372450 |
| 6 11/2 | .358894 | .297549 | .226871 | .000000 | -.200482 | -.349771 | -.428206 |
| 6 13/2 | .158309 | .092159 | .017230 | .000000 | -.025835 | -.075202 | -.124378 |

**K = 3/2+ [602]**

| | -0.3 | -0.2 | -0.1 | 0.0 | 0.1 | 0.2 | 0.3 |
|---|---|---|---|---|---|---|---|
| E | 7.997490 | 7.758702 | 7.556213 | 7.515720 | 7.655094 | 7.824413 | 8.001812 |
| L J | | | | | | | |
| 2 3/2 | .433767 | .512098 | .707833 | 1.000000 | .951101 | .915191 | .892929 |
| 2 5/2 | -.464780 | -.442291 | -.350252 | .000000 | .173100 | .233788 | .266178 |
| 4 7/2 | .643778 | .659530 | .588504 | .000000 | -.251559 | -.314809 | -.340421 |
| 4 9/2 | -.308936 | -.228041 | -.106274 | .000000 | -.033333 | -.069629 | -.096953 |
| 6 11/2 | .281829 | .229720 | .119572 | .000000 | .032295 | .060991 | .078957 |
| 6 13/2 | -.080502 | -.044361 | -.011304 | .000000 | .002785 | .009759 | .017492 |

**K = 1/2+ [660]**

| | -0.3 | -0.2 | -0.1 | 0.0 | 0.1 | 0.2 | 0.3 |
|---|---|---|---|---|---|---|---|
| E | 6.390217 | 6.395608 | 6.352204 | 6.260040 | 6.119429 | 5.931101 | 5.697223 |
| a | 6.200874 | 6.659701 | 6.810928 | 7.000000 | 6.917447 | 6.665806 | 6.238789 |
| L J | | | | | | | |
| 0 1/2 | -.107137 | -.033635 | -.004217 | .000000 | .003631 | .024794 | .067229 |
| 2 3/2 | -.013336 | .000285 | .000609 | .000000 | -.001500 | -.013431 | -.044395 |
| 2 5/2 | .235937 | .115896 | .030638 | .000000 | .029994 | .110162 | .215351 |
| 4 7/2 | .000200 | -.004991 | -.002064 | .000000 | -.003180 | -.016249 | -.044468 |
| 4 9/2 | -.502823 | -.373272 | -.190477 | .000000 | .197375 | .369001 | .495202 |
| 6 11/2 | .005005 | .008062 | .005361 | .000000 | -.007990 | -.019827 | -.037873 |
| 6 13/2 | .824513 | .919791 | .979600 | 1.000000 | .979824 | .922090 | .835761 |

| | β = | -0.3 | -0.2 | -0.1 | 0.0 | 0.1 | 0.2 | 0.3 |
|---|---|---|---|---|---|---|---|---|
| **K = 1/2+[651]** | | | | | | | | |
| | E | 6.838260 | 6.930934 | 6.932431 | 6.852400 | 6.704073 | 6.484872 | 6.207234 |
| | a | .721748 | 2.960655 | 4.743160 | 5.000000 | 4.581716 | 1.759158 | -.450145 |
| L | J | | | | | | | |
| 0 | 1/2 | .533750 | .375446 | .101124 | .000000 | .076000 | .205682 | .262991 |
| 2 | 3/2 | .335695 | .151919 | .004361 | .000000 | -.044895 | -.212180 | -.346848 |
| 2 | 5/2 | -.459131 | -.541251 | -.353882 | .000000 | .340063 | .476915 | .422584 |
| 4 | 7/2 | -.346960 | -.180678 | -.009301 | .000000 | -.072081 | -.305803 | -.460002 |
| 4 | 9/2 | .231238 | .592069 | .907057 | 1.000000 | .904130 | .585450 | .280828 |
| 6 | 11/2 | .313830 | .240805 | .060717 | .000000 | -.127357 | -.388078 | -.460165 |
| 6 | 13/2 | .345365 | .319073 | .194927 | .000000 | -.194160 | -.313616 | -.360190 |
| **K = 1/2+[640]** | | | | | | | | |
| | E | 6.984034 | 6.998662 | 6.989190 | 6.910040 | 6.759370 | 6.559549 | 6.354802 |
| | a | -1.637837 | -3.949451 | -5.756895 | -6.000000 | -5.591388 | -2.752183 | -.453793 |
| L | J | | | | | | | |
| 0 | 1/2 | -.164790 | .002161 | .015287 | .000000 | -.002903 | .059055 | .160569 |
| 2 | 3/2 | .234600 | .222773 | .076069 | .000000 | .063189 | .109340 | .063326 |
| 2 | 5/2 | .368058 | .178520 | .013480 | .000000 | .038206 | .262216 | .433432 |
| 4 | 7/2 | -.481038 | -.476330 | -.294427 | .000000 | .280621 | .383499 | .354341 |
| 4 | 9/2 | -.303049 | -.276595 | -.059518 | .000000 | .140255 | .412492 | .427278 |
| 6 | 11/2 | .601834 | .770938 | .950366 | 1.000000 | .946410 | .753992 | .605442 |
| 6 | 13/2 | -.311288 | -.144074 | -.016265 | .000000 | -.020687 | -.173423 | -.328115 |
| **K = 1/2+[631]** | | | | | | | | |
| | E | 8.016698 | 7.767879 | 7.554024 | 7.265720 | 7.092276 | 6.897853 | 6.748990 |
| | a | .028850 | -.001101 | -.012042 | 3.000000 | -.054412 | -.847004 | -.957601 |
| L | J | | | | | | | |
| 0 | 1/2 | .296273 | .384293 | .576703 | .000000 | .364301 | .377403 | .335676 |
| 2 | 3/2 | .321924 | .332127 | .396435 | .000000 | -.429901 | -.542492 | -.546420 |
| 2 | 5/2 | .580639 | .591736 | .617721 | 1.000000 | .574025 | .264694 | .077183 |
| 4 | 7/2 | .534751 | .528557 | .462847 | .000000 | -.490247 | -.381321 | -.271997 |
| 4 | 9/2 | .343602 | .266773 | .130826 | .000000 | -.272189 | -.413140 | -.458731 |
| 6 | 11/2 | .245189 | .197329 | .103480 | .000000 | .193119 | .404080 | .506561 |
| 6 | 13/2 | .085478 | .048790 | .012929 | .000000 | .035233 | .117627 | .204373 |
| **K = 1/2+[620]** | | | | | | | | |
| | E | 7.620398 | 7.479776 | 7.375357 | 7.302400 | 7.183091 | 7.115439 | 7.076683 |
| | a | -.738070 | -.995220 | -1.692315 | -4.000000 | -.886097 | -.024165 | .247429 |
| L | J | | | | | | | |
| 0 | 1/2 | .372859 | .323110 | .163391 | .000000 | .205496 | .332757 | .384710 |
| 2 | 3/2 | -.419895 | -.434947 | -.421524 | .000000 | .236409 | .197704 | .169806 |
| 2 | 5/2 | -.249651 | -.375461 | -.471006 | .000000 | .598600 | .559894 | .476719 |
| 4 | 7/2 | .431763 | .518540 | .689262 | 1.000000 | .677116 | .545658 | .492942 |
| 4 | 9/2 | -.455713 | -.380404 | -.203359 | .000000 | -.201446 | -.374680 | -.459114 |
| 6 | 11/2 | .447373 | .377760 | .238099 | .000000 | -.209749 | -.303588 | -.350991 |
| 6 | 13/2 | -.167638 | -.095612 | -.025404 | .000000 | .022343 | .080068 | .137593 |
| **K = 1/2+[600]** | | | | | | | | |
| | E | 7.268972 | 7.237924 | 7.282523 | 7.500000 | 7.643541 | 7.812370 | 7.989588 |
| | a | .030932 | .121365 | .812949 | 1.000000 | -.017920 | .125717 | .229029 |

| L | J | β = −0.3 | −0.2 | −0.1 | 0.0 | 0.1 | 0.2 | 0.3 |
|---|---|---|---|---|---|---|---|---|
| 0 | 1/2 | −.486283 | .613151 | −.611316 | 1.000000 | .770289 | .755044 | .743224 |
| 2 | 3/2 | −.417551 | .441641 | −.345056 | .000000 | .556099 | .514478 | .491598 |
| 2 | 5/2 | −.107710 | −.139779 | .646228 | .000000 | −.234537 | −.314348 | −.353986 |
| 4 | 7/2 | .042918 | −.179881 | .335521 | .000000 | −.200276 | −.239262 | −.252657 |
| 4 | 9/2 | .494906 | −.448750 | .266761 | .000000 | .036459 | .080371 | .111053 |
| 6 | 11/2 | .513253 | −.395224 | .151148 | .000000 | .026320 | .050878 | .064198 |
| 6 | 13/2 | .259569 | −.139727 | .034809 | .000000 | −.002990 | −.010539 | −.018795 |

**K = 1/2+[611]**

| | | −0.3 | −0.2 | −0.1 | 0.0 | 0.1 | 0.2 | 0.3 |
|---|---|---|---|---|---|---|---|---|
| E | | 8.474708 | 8.120122 | 7.782913 | 7.515720 | 7.442219 | 7.480491 | 7.544833 |
| a | | −.606498 | −.795969 | −1.011766 | −2.000000 | −.949346 | −.927330 | −.853709 |
| L | J | | | | | | | |
| 0 | 1/2 | −.461740 | −.479446 | −.606490 | .000000 | −.475303 | −.361028 | −.295755 |
| 2 | 3/2 | .614653 | .657844 | .735042 | 1.000000 | .666362 | .587172 | .551940 |
| 2 | 5/2 | −.432432 | −.388678 | −.592235 | .000000 | .373045 | .457493 | .478183 |
| 4 | 7/2 | .417135 | .399764 | .335674 | .000000 | −.420842 | −.508679 | −.529201 |
| 4 | 9/2 | −.176607 | −.127637 | −.060071 | .000000 | −.088170 | −.182779 | −.244522 |
| 6 | 11/2 | .124871 | .098765 | .052967 | .000000 | .077026 | .149879 | .190612 |
| 6 | 13/2 | −.034877 | −.019338 | −.005277 | .000000 | .006385 | .031007 | .055260 |

**N = 6 $\mu_6 = 0.62$ $\kappa = 0.05$ odd proton nuclei**

**K = 13/2+[606]**

| | | −0.3 | −0.2 | −0.1 | 0.0 | 0.1 | 0.2 | 0.3 |
|---|---|---|---|---|---|---|---|---|
| E | | 5.330295 | 5.519530 | 5.708765 | 5.898000 | 6.087235 | 6.276470 | 6.465705 |
| L | J | | | | | | | |
| 6 | 13/2 | 1.000000 | 1.000000 | 1.000000 | 1.000000 | 1.000000 | 1.000000 | 1.000000 |

**K = 11/2+[615]**

| | | −0.3 | −0.2 | −0.1 | 0.0 | 0.1 | 0.2 | 0.3 |
|---|---|---|---|---|---|---|---|---|
| E | | 5.578798 | 5.689072 | 5.794991 | 5.898000 | 5.999026 | 6.098665 | 6.197306 |
| L | J | | | | | | | |
| 6 | 11/2 | .175888 | .101342 | .044108 | .000000 | −.034473 | −.061886 | −.084066 |
| 6 | 13/2 | .984410 | .994852 | .999027 | 1.000000 | .999406 | .998083 | .996460 |

**K = 11/2+[606]**

| | | −0.3 | −0.2 | −0.1 | 0.0 | 0.1 | 0.2 | 0.3 |
|---|---|---|---|---|---|---|---|---|
| E | | 6.015645 | 6.189223 | 6.367157 | 6.548000 | 6.730826 | 6.915040 | 7.100251 |
| L | J | | | | | | | |
| 6 | 11/2 | .984410 | .994852 | .999027 | 1.000000 | .999406 | .998083 | .996460 |
| 6 | 13/2 | −.175888 | −.101342 | −.044108 | .000000 | .034473 | .061886 | .084066 |

**K = 9/2+[624]**

| | | −0.3 | −0.2 | −0.1 | 0.0 | 0.1 | 0.2 | 0.3 |
|---|---|---|---|---|---|---|---|---|
| E | | 5.749482 | 5.816458 | 5.863850 | 5.898000 | 5.923221 | 5.942317 | 5.957119 |
| L | J | | | | | | | |
| 4 | 9/2 | −.324263 | −.181954 | −.075949 | .000000 | .054694 | .094870 | .125085 |
| 6 | 11/2 | .105406 | .082965 | .043567 | .000000 | −.041037 | −.077152 | −.108043 |
| 6 | 13/2 | .940076 | .979801 | .996159 | 1.000000 | .997660 | .992495 | .986246 |

**K = 9/2+[615]**

| L | J | β = −0.3 | −0.2 | −0.1 | 0.0 | 0.1 | 0.2 | 0.3 |
|---|---|---|---|---|---|---|---|---|
| E | | 6.163984 | 6.316308 | 6.456523 | 6.548000 | 6.630156 | 6.713685 | 6.799197 |
| 4 | 9/2 | .820982 | .797286 | .391695 | .000000 | −.085429 | −.108011 | −.112768 |
| 6 | 11/2 | .525045 | .595655 | .920037 | 1.000000 | .995299 | .990309 | .986063 |
| 6 | 13/2 | .224313 | .097623 | −.010375 | .000000 | .045623 | .087306 | .122325 |

**K = 9/2+[604]**

| L | J | β = −0.3 | −0.2 | −0.1 | 0.0 | 0.1 | 0.2 | 0.3 |
|---|---|---|---|---|---|---|---|---|
| E | | 6.360977 | 6.425530 | 6.521774 | 6.680000 | 6.856475 | 7.037703 | 7.221241 |
| 4 | 9/2 | −.469938 | −.575524 | .916955 | 1.000000 | .994842 | .989613 | .985717 |
| 6 | 11/2 | .844522 | .798944 | −.389403 | .000000 | .087724 | .115483 | .126518 |
| 6 | 13/2 | −.256789 | −.174529 | .086941 | .000000 | −.050931 | −.085617 | −.111157 |

**K = 7/2+[633]**

| L | J | β = −0.3 | −0.2 | −0.1 | 0.0 | 0.1 | 0.2 | 0.3 |
|---|---|---|---|---|---|---|---|---|
| E | | 5.887281 | 5.915030 | 5.917518 | 5.898000 | 5.860765 | 5.809993 | 5.749157 |
| 4 | 7/2 | −.077437 | −.029779 | −.006259 | .000000 | −.004195 | −.013636 | −.025035 |
| 4 | 9/2 | −.356303 | −.232908 | −.109632 | .000000 | .090208 | .161323 | .216400 |
| 6 | 11/2 | .092655 | .067550 | .036868 | .000000 | −.039822 | −.078814 | −.114510 |
| 6 | 13/2 | .926535 | .969693 | .993269 | 1.000000 | .995118 | .983655 | .969243 |

**K = 7/2+[624]**

| L | J | β = −0.3 | −0.2 | −0.1 | 0.0 | 0.1 | 0.2 | 0.3 |
|---|---|---|---|---|---|---|---|---|
| E | | 6.419588 | 6.494288 | 6.533702 | 6.548000 | 6.545108 | 6.536628 | 6.527580 |
| 4 | 7/2 | −.556630 | −.264404 | −.110652 | .000000 | .083984 | .138535 | .174016 |
| 4 | 9/2 | .096718 | .284586 | .205381 | .000000 | −.111838 | −.148214 | −.152785 |
| 6 | 11/2 | .820041 | .921458 | .972304 | 1.000000 | .988905 | .973640 | .960850 |
| 6 | 13/2 | −.091333 | −.003956 | −.014118 | .000000 | .050065 | .104239 | .152125 |

**K = 7/2+[613]**

| L | J | β = −0.3 | −0.2 | −0.1 | 0.0 | 0.1 | 0.2 | 0.3 |
|---|---|---|---|---|---|---|---|---|
| E | | 6.519429 | 6.579813 | 6.630089 | 6.680000 | 6.744327 | 6.818687 | 6.898991 |
| 4 | 7/2 | .425991 | .383463 | .146045 | .000000 | −.060633 | −.092490 | −.114225 |
| 4 | 9/2 | .812500 | .878890 | .964973 | 1.000000 | .987145 | .969031 | .953446 |
| 6 | 11/2 | .229535 | −.160403 | −.185552 | .000000 | .121086 | .176312 | .202673 |
| 6 | 13/2 | .325099 | .234048 | .114316 | .000000 | −.084895 | −.146080 | −.191879 |

**K = 7/2+[604]**

| L | J | β = −0.3 | −0.2 | −0.1 | 0.0 | 0.1 | 0.2 | 0.3 |
|---|---|---|---|---|---|---|---|---|
| E | | 6.861997 | 6.888400 | 6.985457 | 7.130000 | 7.295035 | 7.469162 | 7.647977 |
| 4 | 7/2 | .709012 | .884398 | .983050 | 1.000000 | .994612 | .985935 | .977775 |
| 4 | 9/2 | −.451152 | −.303837 | −.120940 | .000000 | .070002 | .113961 | .144114 |
| 6 | 11/2 | .516005 | .347307 | .137243 | .000000 | −.076289 | −.121358 | −.150260 |
| 6 | 13/2 | −.165837 | −.070012 | −.012249 | .000000 | −.005205 | −.014746 | −.024673 |

**K = 5/2+[642]**

| L | J | β = −0.3 | −0.2 | −0.1 | 0.0 | 0.1 | 0.2 | 0.3 |
|---|---|---|---|---|---|---|---|---|
| E | | 5.984569 | 5.986760 | 5.957056 | 5.898000 | 5.812653 | 5.704862 | 5.578850 |

| L | J | β = −0.3 | −0.2 | −0.1 | 0.0 | 0.1 | 0.2 | 0.3 |
|---|---|---|---|---|---|---|---|---|
| 2 | 5/2 | .122929 | .048462 | .010450 | .000000 | .007265 | .023696 | .043266 |
| 4 | 7/2 | -.039682 | -.022045 | -.006076 | .000000 | -.005871 | -.021411 | -.042615 |
| 4 | 9/2 | -.390975 | -.265733 | -.131641 | .000000 | .119115 | .219461 | .299552 |
| 6 | 11/2 | .059067 | .048009 | .027493 | .000000 | -.032997 | -.069115 | -.105289 |
| 6 | 13/2 | .909375 | .961377 | .990842 | 1.000000 | .992288 | .972646 | .946306 |

**K = 5/2+[633]**

| L | J | −0.3 | −0.2 | −0.1 | 0.0 | 0.1 | 0.2 | 0.3 |
|---|---|---|---|---|---|---|---|---|
| E | | 6.515646 | 6.583467 | 6.587114 | 6.548000 | 6.477951 | 6.388734 | 6.291147 |
| 2 | 5/2 | -.517114 | -.223799 | -.041184 | .000000 | -.020088 | -.049634 | -.069513 |
| 4 | 7/2 | -.390045 | -.315446 | -.161543 | .000000 | .138185 | .238215 | .304571 |
| 4 | 9/2 | .344372 | .272828 | .135442 | .000000 | -.117815 | -.170456 | -.174301 |
| 6 | 11/2 | .660984 | .880177 | .976611 | 1.000000 | .982008 | .948222 | .917440 |
| 6 | 13/2 | .158008 | .035506 | -.009660 | .000000 | .047763 | .112293 | .174146 |

**K = 5/2+[622]**

| L | J | −0.3 | −0.2 | −0.1 | 0.0 | 0.1 | 0.2 | 0.3 |
|---|---|---|---|---|---|---|---|---|
| E | | 6.650994 | 6.686849 | 6.700355 | 6.680000 | 6.650602 | 6.625785 | 6.607117 |
| 2 | 5/2 | -.528990 | -.478915 | -.218462 | .000000 | .131667 | .207580 | .254570 |
| 4 | 7/2 | .303270 | .167922 | .091082 | .000000 | -.063282 | -.093096 | -.110492 |
| 4 | 9/2 | .466274 | .763281 | .954249 | 1.000000 | .973541 | .926182 | .881823 |
| 6 | 11/2 | -.554610 | -.308446 | -.125171 | .000000 | .133930 | .224448 | .273859 |
| 6 | 13/2 | .321236 | .254373 | .133115 | .000000 | -.113750 | -.200135 | -.265284 |

**K = 5/2+[613]**

| L | J | −0.3 | −0.2 | −0.1 | 0.0 | 0.1 | 0.2 | 0.3 |
|---|---|---|---|---|---|---|---|---|
| E | | 7.348163 | 7.237587 | 7.170192 | 7.130000 | 7.158510 | 7.208353 | 7.272767 |
| 2 | 5/2 | -.411337 | -.463928 | -.600256 | .000000 | -.196742 | -.208867 | -.198196 |
| 4 | 7/2 | .713056 | .774266 | .765819 | 1.000000 | .966398 | .936681 | .912514 |
| 4 | 9/2 | -.429266 | -.332857 | -.191264 | .000000 | .105738 | .185012 | .236873 |
| 6 | 11/2 | .351460 | .265289 | .127711 | .000000 | -.126853 | -.209330 | -.262689 |
| 6 | 13/2 | -.120667 | -.064111 | -.017928 | .000000 | -.009753 | -.030912 | -.054054 |

**K = 5/2+[602]**

| L | J | −0.3 | −0.2 | −0.1 | 0.0 | 0.1 | 0.2 | 0.3 |
|---|---|---|---|---|---|---|---|---|
| E | | 6.970627 | 6.975337 | 7.055283 | 7.214000 | 7.370284 | 7.542267 | 7.720118 |
| 2 | 5/2 | .518125 | .709208 | .768220 | 1.000000 | .971339 | .954075 | .942978 |
| 4 | 7/2 | .495852 | .521844 | .615704 | .000000 | .207221 | .238239 | .246029 |
| 4 | 9/2 | .571721 | .401944 | .130970 | .000000 | -.113876 | -.175326 | -.214867 |
| 6 | 11/2 | .358461 | .239721 | .116174 | .000000 | -.023292 | -.043614 | -.056683 |
| 6 | 13/2 | .174119 | .075345 | .009850 | .000000 | .007010 | .018461 | .029674 |

**K = 3/2+[651]**

| L | J | −0.3 | −0.2 | −0.1 | 0.0 | 0.1 | 0.2 | 0.3 |
|---|---|---|---|---|---|---|---|---|
| E | | 6.049492 | 6.033875 | 5.983111 | 5.898000 | 5.779860 | 5.630763 | 5.453815 |
| 2 | 3/2 | .039052 | .010934 | .001249 | .000000 | -.000946 | -.006274 | -.017070 |
| 2 | 5/2 | .135349 | .062069 | .015376 | .000000 | .013490 | .048021 | .093513 |
| 4 | 7/2 | -.032988 | -.015429 | -.004275 | .000000 | -.005186 | -.021604 | -.048224 |
| 4 | 9/2 | -.406091 | -.284304 | -.145162 | .000000 | .139975 | .264785 | .367591 |

| β = | | -0.3 | -0.2 | -0.1 | 0.0 | 0.1 | 0.2 | 0.3 |
|---|---|---|---|---|---|---|---|---|
| 6 | 11/2 | .038507 | .029199 | .016884 | .000000 | -.021808 | -.048493 | -.078898 |
| 6 | 13/2 | .901485 | .956090 | .989134 | 1.000000 | .989809 | .961626 | .920484 |

K = 3/2+[642]

| | | | | | | | | |
|---|---|---|---|---|---|---|---|---|
| E | | 6.623404 | 6.649647 | 6.621952 | 6.548000 | 6.431324 | 6.277768 | 6.099707 |
| L | J | | | | | | | |
| 2 | 3/2 | .331560 | .120639 | .026117 | .000000 | .020039 | .063903 | .107192 |
| 2 | 5/2 | .035728 | -.051068 | -.024094 | .000000 | -.028442 | -.090975 | -.140662 |
| 4 | 7/2 | -.516998 | -.368994 | -.189349 | .000000 | .178843 | .322451 | .418662 |
| 4 | 9/2 | -.004895 | .086977 | .073893 | .000000 | -.098056 | -.170248 | -.178989 |
| 6 | 11/2 | .784824 | .916013 | .978460 | 1.000000 | .977670 | .918462 | .853220 |
| 6 | 13/2 | -.074375 | -.006130 | -.006335 | .000000 | .036751 | .105398 | .182822 |

K = 3/2+[631]

| | | | | | | | | |
|---|---|---|---|---|---|---|---|---|
| E | | 6.738876 | 6.764690 | 6.745929 | 6.680000 | 6.579897 | 6.465986 | 6.352482 |
| L | J | | | | | | | |
| 2 | 3/2 | .346480 | -.202250 | -.044611 | .000000 | -.023527 | -.059445 | -.086417 |
| 2 | 5/2 | .576164 | -.474181 | -.259141 | .000000 | .217056 | .353183 | .428097 |
| 4 | 7/2 | -.037356 | .115727 | .059925 | .000000 | -.047820 | -.058586 | -.048282 |
| 4 | 9/2 | -.599578 | .802936 | .949812 | 1.000000 | .958061 | .864221 | .767118 |
| 6 | 11/2 | -.235344 | -.027584 | -.064385 | .000000 | .116751 | .247973 | .335730 |
| 6 | 13/2 | -.362920 | .274568 | .144834 | .000000 | -.136144 | -.244801 | -.325191 |

K = 3/2+[622]

| | | | | | | | | |
|---|---|---|---|---|---|---|---|---|
| E | | 7.080772 | 7.106705 | 7.161029 | 7.130000 | 7.045060 | 6.970541 | 6.919561 |
| L | J | | | | | | | |
| 2 | 3/2 | .710441 | -.748538 | -.458678 | .000000 | .242190 | .334287 | .372072 |
| 2 | 5/2 | -.100042 | -.071252 | .033821 | .000000 | -.259927 | -.258646 | -.211462 |
| 4 | 7/2 | -.279045 | .503658 | .866842 | 1.000000 | .910821 | .821179 | .754578 |
| 4 | 9/2 | .405630 | -.268731 | -.053382 | .000000 | .128607 | .251853 | .326684 |
| 6 | 11/2 | -.460154 | .322555 | .184809 | .000000 | -.165741 | -.284734 | -.363261 |
| 6 | 13/2 | .176413 | -.068447 | -.007189 | .000000 | -.013292 | -.050161 | -.093681 |

K = 3/2+[611]

| | | | | | | | | |
|---|---|---|---|---|---|---|---|---|
| E | | 7.424062 | 7.302603 | 7.235422 | 7.214000 | 7.216915 | 7.260021 | 7.321682 |
| L | J | | | | | | | |
| 2 | 3/2 | .187403 | .227138 | .379905 | .000000 | -.115369 | -.133310 | -.146172 |
| 2 | 5/2 | .591958 | .698878 | .874060 | 1.000000 | .922642 | .858780 | .816294 |
| 4 | 7/2 | .557656 | .527642 | .176300 | .000000 | .312181 | .383010 | .400638 |
| 4 | 9/2 | .461828 | .377356 | .243687 | .000000 | -.188444 | -.294102 | -.359697 |
| 6 | 11/2 | .275456 | .186215 | .027240 | .000000 | -.047348 | -.099327 | -.133579 |
| 6 | 13/2 | .119685 | .067070 | .021993 | .000000 | .014557 | .040822 | .067544 |

K = 3/2+[602]

| | | | | | | | | |
|---|---|---|---|---|---|---|---|---|
| E | | 7.868951 | 7.644184 | 7.470410 | 7.464000 | 7.597090 | 7.761217 | 7.935196 |
| L | J | | | | | | | |
| 2 | 3/2 | -.478172 | .576647 | .801631 | 1.000000 | .962849 | .928883 | .906061 |
| 2 | 5/2 | .536671 | -.524580 | -.408538 | .000000 | .181841 | .245516 | .277760 |
| 4 | 7/2 | -.584280 | .564034 | .421949 | .000000 | -.196593 | -.266637 | -.300276 |
| 4 | 9/2 | .312652 | -.228663 | -.095355 | .000000 | -.029340 | -.064038 | -.090915 |

| L | J | β = −0.3 | −0.2 | −0.1 | 0.0 | 0.1 | 0.2 | 0.3 |
|---|---|---|---|---|---|---|---|---|
| 6 | 11/2 | −.198908 | .143490 | .057348 | .000000 | .018501 | .040571 | .057217 |
| 6 | 13/2 | .068095 | −.035814 | −.007811 | .000000 | .001968 | .007489 | .014064 |

**K = 1/2+[660]**

| | | −0.3 | −0.2 | −0.1 | 0.0 | 0.1 | 0.2 | 0.3 |
|---|---|---|---|---|---|---|---|---|
| E | | 6.081145 | 6.057181 | 5.996052 | 5.898000 | 5.763216 | 5.591969 | 5.384827 |
| a | | 6.551551 | 6.806068 | 6.952269 | 7.000000 | 6.952320 | 6.806419 | 6.552659 |
| L | J | | | | | | | |
| 0 | 1/2 | −.050707 | −.015218 | −.001876 | .000000 | .001680 | .012172 | .036022 |
| 2 | 3/2 | −.001490 | .001279 | .000368 | .000000 | −.000700 | −.006384 | −.022669 |
| 2 | 5/2 | .148139 | .069416 | .017868 | .000000 | .017599 | .067105 | .139660 |
| 4 | 7/2 | −.006018 | −.004715 | −.001505 | .000000 | −.002108 | −.010213 | −.027731 |
| 4 | 9/2 | −.414354 | −.293107 | −.151634 | .000000 | .151016 | .290666 | .409077 |
| 6 | 11/2 | .010541 | .009544 | .005683 | .000000 | −.007638 | −.017956 | −.032538 |
| 6 | 13/2 | .896463 | .953374 | .988256 | 1.000000 | .988341 | .954146 | .899729 |

**K = 1/2+[651]**

| | | −0.3 | −0.2 | −0.1 | 0.0 | 0.1 | 0.2 | 0.3 |
|---|---|---|---|---|---|---|---|---|
| E | | 6.656980 | 6.678836 | 6.639057 | 6.548000 | 6.407597 | 6.216762 | 5.974720 |
| a | | −4.127188 | −5.538146 | −5.905655 | −6.000000 | −5.890264 | −5.377412 | −4.273120 |
| L | J | | | | | | | |
| 0 | 1/2 | .240051 | .063930 | .007003 | .000000 | −.006373 | −.047667 | −.122800 |
| 2 | 3/2 | .273540 | .132408 | .034061 | .000000 | .033183 | .125354 | .241294 |
| 2 | 5/2 | −.196755 | −.055359 | −.009738 | .000000 | −.016260 | −.088120 | −.191813 |
| 4 | 7/2 | −.489675 | −.380921 | −.201930 | .000000 | .200722 | .375565 | .489644 |
| 4 | 9/2 | .139698 | .062589 | .025061 | .000000 | −.041863 | −.112878 | −.155693 |
| 6 | 11/2 | .748154 | .908914 | .978411 | 1.000000 | .977924 | .903594 | .775873 |
| 6 | 13/2 | .099032 | .013133 | −.001912 | .000000 | .014706 | .063055 | .154709 |

**K = 1/2+[640]**

| | | −0.3 | −0.2 | −0.1 | 0.0 | 0.1 | 0.2 | 0.3 |
|---|---|---|---|---|---|---|---|---|
| E | | 6.780265 | 6.798858 | 6.767853 | 6.680000 | 6.539615 | 6.354494 | 6.148348 |
| a | | 2.583063 | 4.325427 | 4.863990 | 5.000000 | 4.847744 | 4.185695 | 2.912589 |
| L | J | | | | | | | |
| 0 | 1/2 | −.391498 | .236757 | .060654 | .000000 | .048813 | .149470 | .225649 |
| 2 | 3/2 | −.036188 | .023721 | −.005079 | .000000 | −.022311 | −.080371 | −.111262 |
| 2 | 5/2 | .555397 | −.497538 | −.280709 | .000000 | .273734 | .467533 | .542310 |
| 4 | 7/2 | −.139771 | .029140 | .017576 | .000000 | −.020046 | −.012522 | .060941 |
| 4 | 9/2 | −.522372 | .779132 | .945523 | 1.000000 | .946962 | .800424 | .615030 |
| 6 | 11/2 | .338665 | −.095906 | −.023349 | .000000 | .052524 | .189065 | .361270 |
| 6 | 13/2 | −.360349 | .280616 | .150431 | .000000 | −.149303 | −.275739 | −.360708 |

**K = 1/2+[631]**

| | | −0.3 | −0.2 | −0.1 | 0.0 | 0.1 | 0.2 | 0.3 |
|---|---|---|---|---|---|---|---|---|
| E | | 7.116086 | 7.135267 | 7.186670 | 7.130000 | 6.968959 | 6.761867 | 6.584835 |
| a | | −.360895 | −.914457 | −2.809947 | −4.000000 | −3.033429 | −1.998209 | −1.536127 |
| L | J | | | | | | | |
| 0 | 1/2 | −.497677 | .551698 | −.297594 | .000000 | −.173998 | −.288640 | −.289458 |
| 2 | 3/2 | −.514090 | .516816 | −.411111 | .000000 | .395987 | .539861 | .565109 |
| 2 | 5/2 | −.023117 | −.204605 | .245039 | .000000 | −.252123 | −.222633 | −.088506 |
| 4 | 7/2 | .277657 | −.448813 | .801853 | 1.000000 | .837304 | .612739 | .468320 |
| 4 | 9/2 | .453421 | −.307718 | .078065 | .000000 | .117218 | .300803 | .406018 |
| 6 | 11/2 | .414623 | −.292436 | .182386 | .000000 | −.185416 | −.324170 | −.428385 |
| 6 | 13/2 | .181379 | −.070887 | .007308 | .000000 | −.012492 | −.068225 | −.146096 |

**K = 1/2+[620]**

| | β = -0.3 | -0.2 | -0.1 | 0.0 | 0.1 | 0.2 | 0.3 |
|---|---|---|---|---|---|---|---|
| E | 7.469325 | 7.349285 | 7.273406 | 7.214000 | 7.088717 | 7.000218 | 6.945039 |
| a | -.379176 | -.281062 | 1.573526 | 3.000000 | 1.889474 | .920726 | .729989 |

| L | J | -0.3 | -0.2 | -0.1 | 0.0 | 0.1 | 0.2 | 0.3 |
|---|---|---|---|---|---|---|---|---|
| 0 | 1/2 | .454363 | .462222 | -.485679 | .000000 | .340987 | .401705 | .426434 |
| 2 | 3/2 | -.317429 | -.265062 | .065190 | .000000 | -.058131 | .018397 | .046588 |
| 2 | 5/2 | -.288113 | -.433348 | .738604 | 1.000000 | .825932 | .640180 | .529314 |
| 4 | 7/2 | .522490 | .568717 | -.379957 | .000000 | .366302 | .502024 | .506605 |
| 4 | 9/2 | -.451104 | -.373202 | .252107 | .000000 | -.242699 | -.374246 | -.451742 |
| 6 | 11/2 | .338738 | .244839 | -.076269 | .000000 | -.067978 | -.178944 | -.249219 |
| 6 | 13/2 | -.136197 | -.075090 | .024242 | .000000 | .022012 | .065989 | .113932 |

**K = 1/2+[611]**

| | -0.3 | -0.2 | -0.1 | 0.0 | 0.1 | 0.2 | 0.3 |
|---|---|---|---|---|---|---|---|
| E | 8.410002 | 8.062573 | 7.736850 | 7.464000 | 7.381675 | 7.398649 | 7.450143 |
| a | -.275225 | -.393173 | -.512653 | -2.000000 | -1.119619 | -.917133 | -.809908 |

| L | J | -0.3 | -0.2 | -0.1 | 0.0 | 0.1 | 0.2 | 0.3 |
|---|---|---|---|---|---|---|---|---|
| 0 | 1/2 | -.505753 | -.538065 | -.599259 | .000000 | -.395119 | -.311657 | -.258851 |
| 2 | 3/2 | .612921 | .648257 | .696113 | 1.000000 | .780803 | .678564 | .624687 |
| 2 | 5/2 | -.454968 | -.413276 | -.315561 | .000000 | .346360 | .455688 | .483264 |
| 4 | 7/2 | .354309 | .318428 | .230390 | .000000 | -.328164 | -.445616 | -.487368 |
| 4 | 9/2 | -.167103 | -.119231 | -.054534 | .000000 | -.069737 | -.161909 | -.225342 |
| 6 | 11/2 | .085370 | .060118 | .025853 | .000000 | .040981 | .097256 | .137907 |
| 6 | 13/2 | -.028268 | -.015051 | -.003856 | .000000 | .005329 | .022851 | .043510 |

**K = 1/2+[600]**

| | -0.3 | -0.2 | -0.1 | 0.0 | 0.1 | 0.2 | 0.3 |
|---|---|---|---|---|---|---|---|
| E | 7.907164 | 7.676645 | 7.496433 | 7.500000 | 7.621899 | 7.785398 | 7.959120 |
| a | .007869 | -.004658 | -.161531 | 1.000000 | .353774 | .379913 | .423917 |

| L | J | -0.3 | -0.2 | -0.1 | 0.0 | 0.1 | 0.2 | 0.3 |
|---|---|---|---|---|---|---|---|---|
| 0 | 1/2 | .276801 | .363447 | .559206 | 1.000000 | .833611 | .795882 | .774642 |
| 2 | 3/2 | .427951 | .473614 | .583934 | .000000 | .478082 | .474920 | .465974 |
| 2 | 5/2 | .583384 | .586619 | .464358 | .000000 | -.242490 | -.319526 | -.357843 |
| 4 | 7/2 | .514802 | .477241 | .344235 | .000000 | -.127899 | -.180663 | -.205585 |
| 4 | 9/2 | .314506 | .234389 | .098684 | .000000 | .034614 | .073054 | .102237 |
| 6 | 11/2 | .177531 | .124525 | .048824 | .000000 | .012945 | .029723 | .042473 |
| 6 | 13/2 | .066701 | .035629 | .007833 | .000000 | -.002222 | -.008132 | -.015012 |

**N = 7 $\mu$ = 0.434 $\kappa$ = 0.05 odd neuton nuclei**

**K = 15/2-[707]**

| | -0.3 | -0.2 | -0.1 | 0.0 | 0.1 | 0.2 | 0.3 |
|---|---|---|---|---|---|---|---|
| E | 6.272477 | 6.493252 | 6.714026 | 6.934800 | 7.155574 | 7.376348 | 7.597123 |

| L | J | -0.3 | -0.2 | -0.1 | 0.0 | 0.1 | 0.2 | 0.3 |
|---|---|---|---|---|---|---|---|---|
| 7 | 15/2 | 1.000000 | 1.000000 | 1.000000 | 1.000000 | 1.000000 | 1.000000 | 1.000000 |

**K = 13/2-[716]**

| | -0.3 | -0.2 | -0.1 | 0.0 | 0.1 | 0.2 | 0.3 |
|---|---|---|---|---|---|---|---|
| E | 6.527648 | 6.666093 | 6.801503 | 6.934800 | 7.066596 | 7.197298 | 7.327185 |

| L | J | -0.3 | -0.2 | -0.1 | 0.0 | 0.1 | 0.2 | 0.3 |
|---|---|---|---|---|---|---|---|---|
| 7 | 13/2 | .136529 | .079781 | .035266 | .000000 | -.028333 | -.051440 | -.070557 |
| 7 | 15/2 | .990636 | .996812 | .999378 | 1.000000 | .999599 | .998676 | .997508 |

| β = | -0.3 | -0.2 | -0.1 | 0.0 | 0.1 | 0.2 | 0.3 |
|---|---|---|---|---|---|---|---|
| **K = 13/2-[707]** | | | | | | | |
| E | 7.051160 | 7.259646 | 7.471167 | 7.684800 | 7.899935 | 8.116164 | 8.333207 |
| L J | | | | | | | |
| 7 13/2 | .990636 | .996812 | .999378 | 1.000000 | .999599 | .998676 | .997508 |
| 7 15/2 | -.136529 | -.079781 | -.035266 | .000000 | .028333 | .051440 | .070557 |
| **K = 11/2-[725]** | | | | | | | |
| E | 6.691191 | 6.795542 | 6.872853 | 6.934800 | 6.987890 | 7.035485 | 7.079438 |
| L J | | | | | | | |
| 5 11/2 | -.424531 | -.226908 | -.088578 | .000000 | .058086 | .098001 | .126671 |
| 7 13/2 | .077572 | .067640 | .036555 | .000000 | -.034740 | -.065502 | -.092073 |
| 7 15/2 | .902084 | .971564 | .995398 | 1.000000 | .997707 | .993028 | .987662 |
| **K = 11/2-[705]** | | | | | | | |
| E | 7.026822 | 7.200150 | 7.394573 | 7.599000 | 7.821726 | 8.029452 | 8.244865 |
| L J | | | | | | | |
| 5 11/2 | .882242 | .959816 | .989557 | 1.000000 | .772674 | .958709 | .971797 |
| 7 13/2 | .259399 | .184666 | .117312 | .000000 | .634390 | .273892 | .211177 |
| 7 15/2 | .392887 | .211308 | .083750 | .000000 | -.022895 | -.076548 | -.104950 |
| **K = 11/2-[716]** | | | | | | | |
| E | 7.365177 | 7.465969 | 7.572705 | 7.684800 | 7.787454 | 7.910603 | 8.029706 |
| L J | | | | | | | |
| 5 11/2 | -.203523 | -.165122 | -.113711 | .000000 | -.632140 | -.266968 | -.198908 |
| 7 13/2 | .962650 | .980471 | .992422 | 1.000000 | .772232 | .959527 | .973102 |
| 7 15/2 | -.178560 | -.106824 | -.046564 | .000000 | .063692 | .089640 | .116227 |
| **K = 9/2-[734]** | | | | | | | |
| E | 6.839679 | 6.900494 | 6.930565 | 6.934800 | 6.920234 | 6.892812 | 6.856645 |
| L J | | | | | | | |
| 5 9/2 | -.080227 | -.030524 | -.006181 | .000000 | -.003771 | -.011832 | -.021185 |
| 5 11/2 | -.428844 | -.279920 | -.128212 | .000000 | .097406 | .168681 | .220830 |
| 7 13/2 | .076618 | .058542 | .032843 | .000000 | -.035421 | -.069552 | -.100490 |
| 7 15/2 | .896541 | .957751 | .991184 | 1.000000 | .994607 | .983143 | .969890 |
| **K = 9/2-[725]** | | | | | | | |
| E | 7.375291 | 7.434656 | 7.509473 | 7.599000 | 7.684061 | 7.721364 | 7.746815 |
| L J | | | | | | | |
| 5 9/2 | .369508 | .137782 | .061551 | .000000 | -.089655 | .152628 | .182465 |
| 5 11/2 | .822800 | .932903 | .980623 | 1.000000 | .817866 | -.412161 | -.290849 |
| 7 13/2 | .111854 | .201564 | .139859 | .000000 | -.559451 | .887977 | .924425 |
| 7 15/2 | .417077 | .264728 | .122595 | .000000 | -.100360 | .135372 | .165987 |
| **K = 9/2-[714]** | | | | | | | |
| E | 7.422227 | 7.565060 | 7.645528 | 7.684800 | 7.728390 | 7.825128 | 7.938786 |
| L J | | | | | | | |
| 5 9/2 | .661118 | -.504240 | -.183733 | .000000 | .039315 | -.023111 | -.056354 |

| β = | | -0.3 | -0.2 | -0.1 | 0.0 | 0.1 | 0.2 | 0.3 |
|---|---|---|---|---|---|---|---|---|
| 5 | 11/2 | -.209408 | -.083812 | -.121275 | .000000 | .564740 | .891168 | .924184 |
| 7 | 13/2 | -.720173 | .854463 | .974229 | 1.000000 | .823928 | .436261 | .333692 |
| 7 | 15/2 | .020539 | -.092794 | -.049114 | .000000 | -.025816 | -.122316 | -.177080 |

**K = 9/2-[705]**

| | | | | | | | | |
|---|---|---|---|---|---|---|---|---|
| E | | 7.784229 | 7.836607 | 7.966642 | 8.149000 | 8.350306 | 8.559079 | 8.771528 |
| L | J | | | | | | | |
| 5 | 9/2 | .648036 | .851954 | .981028 | 1.000000 | .995189 | .987943 | .981367 |
| 5 | 11/2 | -.308613 | -.210508 | -.085046 | .000000 | .051739 | .086542 | .111915 |
| 7 | 13/2 | .680418 | .475225 | .173891 | .000000 | -.083083 | -.127812 | -.154885 |
| 7 | 15/2 | -.147778 | -.063420 | -.010645 | .000000 | -.004253 | -.012000 | -.020093 |

**K = 7/2-[743]**

| | | | | | | | | |
|---|---|---|---|---|---|---|---|---|
| E | | 6.945897 | 6.981269 | 6.975771 | 6.934800 | 6.864478 | 6.771646 | 6.662461 |
| L | J | | | | | | | |
| 3 | 7/2 | .180859 | .067321 | .013453 | .000000 | .007913 | .024012 | .041399 |
| 5 | 9/2 | -.043237 | -.025033 | -.006770 | .000000 | -.005849 | -.020105 | -.038135 |
| 5 | 11/2 | -.463249 | -.317416 | -.155300 | .000000 | .131562 | .233740 | .309620 |
| 7 | 13/2 | .052080 | .045499 | .026871 | .000000 | -.032072 | -.065791 | -.098180 |
| 7 | 15/2 | .864933 | .944467 | .987387 | 1.000000 | .990740 | .969565 | .944102 |

**K = 7/2-[734]**

| | | | | | | | | |
|---|---|---|---|---|---|---|---|---|
| E | | 7.392627 | 7.523434 | 7.585802 | 7.599000 | 7.593387 | 7.550425 | 7.487058 |
| L | J | | | | | | | |
| 3 | 7/2 | .776880 | -.594474 | -.238041 | .000000 | .116076 | -.120752 | -.110171 |
| 5 | 9/2 | .189158 | -.058438 | .028317 | .000000 | -.114615 | .246636 | .310608 |
| 5 | 11/2 | -.414354 | .722609 | .949008 | 1.000000 | .893262 | -.538301 | -.359215 |
| 7 | 13/2 | -.243303 | .215227 | .140547 | .000000 | -.397193 | .773679 | .844176 |
| 7 | 15/2 | -.360265 | .273310 | .148876 | .000000 | -.133080 | .190376 | .222971 |

**K = 7/2-[723]**

| | | | | | | | | |
|---|---|---|---|---|---|---|---|---|
| E | | 7.620776 | 7.676049 | 7.701788 | 7.684800 | 7.649734 | 7.643163 | 7.657526 |
| L | J | | | | | | | |
| 3 | 7/2 | .137148 | .020456 | .008312 | .000000 | .047720 | .155628 | .214736 |
| 5 | 9/2 | -.583878 | -.460648 | -.224090 | .000000 | .105182 | .045850 | -.006981 |
| 5 | 11/2 | -.238018 | -.224090 | -.126653 | .000000 | .406509 | .777979 | .834716 |
| 7 | 13/2 | .728752 | .848707 | .965083 | 1.000000 | .905988 | .588013 | .448334 |
| 7 | 15/2 | -.229225 | -.129866 | -.047834 | .000000 | -.024413 | -.150556 | -.236822 |

**K = 7/2-[703]**

| | | | | | | | | |
|---|---|---|---|---|---|---|---|---|
| E | | 7.811198 | 7.819077 | 7.920065 | 8.089600 | 8.295815 | 8.501324 | 8.712725 |
| L | J | | | | | | | |
| 3 | 7/2 | .527704 | .758502 | .944658 | 1.000000 | .888209 | .930166 | .934959 |
| 5 | 9/2 | .252871 | .256535 | .234582 | .000000 | .447504 | .327319 | .288334 |
| 5 | 11/2 | .667446 | .519527 | .215521 | .000000 | -.084313 | -.150951 | -.191774 |
| 7 | 13/2 | .395504 | .275584 | .075636 | .000000 | -.060777 | -.068015 | -.072542 |
| 7 | 15/2 | .235960 | .114060 | .020577 | .000000 | .004776 | .015527 | .025997 |

**K = 7/2-[714]**

| | | | | | | | | |
|---|---|---|---|---|---|---|---|---|
| E | | 8.213615 | 8.141979 | 8.116078 | 8.149000 | 8.211482 | 8.306033 | 8.410516 |

| β = | | -0.3 | -0.2 | -0.1 | 0.0 | 0.1 | 0.2 | 0.3 |
|---|---|---|---|---|---|---|---|---|
| L | J | | | | | | | |
| 3 | 7/2 | -.257812 | -.257534 | -.225185 | .000000 | -.441895 | -.308890 | -.256688 |
| 5 | 9/2 | .746654 | .847320 | .945467 | 1.000000 | .880628 | .910783 | .904922 |
| 5 | 11/2 | -.334027 | -.238660 | -.113027 | .000000 | .111425 | .166013 | .203890 |
| 7 | 13/2 | .500587 | .394154 | .205956 | .000000 | -.129234 | -.216118 | -.267322 |
| 7 | 15/2 | -.117810 | -.058382 | -.013832 | .000000 | -.010252 | -.028151 | -.046858 |

**K = 5/2-[752]**

| | | | | | | | | |
|---|---|---|---|---|---|---|---|---|
| E | | 7.028057 | 7.040880 | 7.009181 | 6.934800 | 6.821516 | 6.674955 | 6.501935 |
| L | J | | | | | | | |
| 3 | 5/2 | .050099 | .013626 | .001464 | .000000 | -.000916 | -.005461 | -.013494 |
| 3 | 7/2 | .188010 | .086286 | .020681 | .000000 | .015799 | .051725 | .093297 |
| 5 | 9/2 | -.041704 | -.020610 | -.005805 | .000000 | -.006388 | -.024378 | -.049849 |
| 5 | 11/2 | -.475974 | -.339974 | -.173659 | .000000 | .159468 | .290256 | .388476 |
| 7 | 13/2 | .040551 | .032926 | .019770 | .000000 | -.025512 | -.054835 | -.085230 |
| 7 | 15/2 | .855691 | .935563 | .984372 | 1.000000 | .986726 | .953648 | .911291 |

**K = 5/2-[743]**

| | | | | | | | | |
|---|---|---|---|---|---|---|---|---|
| E | | 7.598761 | 7.641090 | 7.642460 | 7.599000 | 7.518852 | 7.401210 | 7.254143 |
| L | J | | | | | | | |
| 3 | 5/2 | -.452349 | -.134394 | -.029987 | .000000 | -.022192 | -.067636 | .100671 |
| 3 | 7/2 | -.579443 | -.525328 | -.283500 | .000000 | .204237 | .257530 | -.236660 |
| 5 | 9/2 | .214091 | .002175 | .015922 | .000000 | -.111420 | -.295904 | .407031 |
| 5 | 11/2 | .471837 | .760468 | .935935 | 1.000000 | .920522 | .633935 | -.400999 |
| 7 | 13/2 | -.077214 | .156257 | .118306 | .000000 | -.269261 | -.614070 | .724694 |
| 7 | 15/2 | .430348 | .321303 | .168832 | .000000 | -.159742 | -.250176 | .286706 |

**K = 5/2-[732]**

| | | | | | | | | |
|---|---|---|---|---|---|---|---|---|
| E | | 7.650378 | 7.733698 | 7.741925 | 7.684800 | 7.586808 | 7.488015 | 7.405768 |
| L | J | | | | | | | |
| 3 | 5/2 | -.340138 | .249978 | .047797 | .000000 | .013642 | .004311 | -.021632 |
| 3 | 7/2 | .255420 | .042633 | .020906 | .000000 | .053639 | .220852 | .339520 |
| 5 | 9/2 | .537476 | -.476222 | -.249019 | .000000 | .172059 | .166469 | .105429 |
| 5 | 11/2 | -.256045 | -.066556 | -.102264 | .000000 | .281250 | .621161 | .701264 |
| 7 | 13/2 | -.670874 | .836264 | .960866 | 1.000000 | .942230 | .716558 | .552136 |
| 7 | 15/2 | -.120642 | -.071681 | -.039318 | .000000 | -.020825 | -.155555 | -.276617 |

**K = 5/2-[712]**

| | | | | | | | | |
|---|---|---|---|---|---|---|---|---|
| E | | 8.244292 | 8.142977 | 8.080933 | 8.089600 | 8.169563 | 8.254667 | 8.356155 |
| L | J | | | | | | | |
| 3 | 5/2 | .031493 | -.073347 | .194342 | .000000 | .035836 | -.019468 | -.053641 |
| 3 | 7/2 | .633141 | .733048 | .923602 | 1.000000 | .756473 | .831053 | .828818 |
| 5 | 9/2 | .426866 | .458261 | .179620 | .000000 | .629425 | .475434 | .424390 |
| 5 | 11/2 | .526454 | .400336 | .271887 | .000000 | -.124653 | -.245979 | -.318224 |
| 7 | 13/2 | .337974 | .284153 | .046886 | .000000 | -.121114 | -.145985 | -.159404 |
| 7 | 15/2 | .156669 | .079033 | .028390 | .000000 | .009010 | .033438 | .058315 |

**K = 5/2-[723]**

| | | | | | | | | |
|---|---|---|---|---|---|---|---|---|
| E | | 7.987468 | 8.036511 | 8.144749 | 8.149000 | 8.086189 | 8.068012 | 8.065490 |

| β = | -0.3 | -0.2 | -0.1 | 0.0 | 0.1 | 0.2 | 0.3 |
|---|---|---|---|---|---|---|---|
| L J | | | | | | | |
| 3 5/2 | .714611 | .802827 | -.565243 | .000000 | .217212 | .300411 | .336476 |
| 3 7/2 | -.059069 | .202902 | -.026369 | .000000 | -.608634 | -.406245 | -.309174 |
| 5 9/2 | -.019366 | -.263257 | .790090 | 1.000000 | .725130 | .769983 | .747222 |
| 5 11/2 | .352100 | .319044 | -.066269 | .000000 | .179542 | .256436 | .302313 |
| 7 13/2 | -.569412 | -.366910 | .225982 | .000000 | -.154910 | -.288895 | -.366239 |
| 7 15/2 | .193034 | .092645 | -.010175 | .000000 | -.018380 | -.051224 | -.085618 |

**K = 5/2-[703]**

| | | | | | | | |
|---|---|---|---|---|---|---|---|
| E | 8.671697 | 8.490880 | 8.372169 | 8.439600 | 8.619254 | 8.820705 | 9.029457 |
| L J | | | | | | | |
| 3 5/2 | .406837 | .518992 | .799716 | 1.000000 | .975118 | .951184 | .934410 |
| 3 7/2 | -.399127 | -.369105 | -.255004 | .000000 | .111687 | .162921 | .193615 |
| 5 9/2 | .693509 | .702478 | .530280 | .000000 | -.189607 | -.255387 | -.286840 |
| 5 11/2 | -.290054 | -.199043 | -.071386 | .000000 | -.018249 | -.042096 | -.062081 |
| 7 13/2 | .322280 | .244533 | .095303 | .000000 | .019624 | .041026 | .056205 |
| 7 15/2 | -.078938 | -.038977 | -.007212 | .000000 | .001346 | .005253 | .010045 |

**K = 3/2-[761]**

| | | | | | | | |
|---|---|---|---|---|---|---|---|
| E | 7.079986 | 7.079982 | 7.031238 | 6.934800 | 6.792228 | 6.606427 | 6.382385 |
| L J | | | | | | | |
| 1 3/2 | -.084290 | -.023943 | -.002722 | .000000 | .001929 | .012053 | .030522 |
| 3 5/2 | .011515 | .006382 | .001080 | .000000 | -.001248 | -.009184 | -.026419 |
| 3 7/2 | .210555 | .099888 | .025534 | .000000 | .023120 | .081600 | .155119 |
| 5 9/2 | -.019109 | -.012380 | -.003829 | .000000 | -.005015 | -.021507 | -.048942 |
| 5 11/2 | -.486035 | -.353865 | -.185139 | .000000 | .179482 | .333534 | .450138 |
| 7 13/2 | .021708 | .019618 | .012060 | .000000 | -.016436 | -.037057 | -.060951 |
| 7 15/2 | .843423 | .929328 | .982295 | 1.000000 | .983337 | .938100 | .874971 |

**K = 3/2-[752]**

| | | | | | | | |
|---|---|---|---|---|---|---|---|
| E | 7.607065 | 7.689086 | 7.678574 | 7.599000 | 7.464352 | 7.280732 | 7.054727 |
| L J | | | | | | | |
| 1 3/2 | .568583 | .308854 | .068553 | .000000 | .039293 | .100070 | -.128549 |
| 3 5/2 | .185903 | .028533 | -.012399 | .000000 | -.030743 | -.124745 | .216554 |
| 3 7/2 | -.534422 | -.541231 | -.314226 | .000000 | .277436 | .403922 | -.379320 |
| 5 9/2 | -.228668 | -.073420 | .004239 | .000000 | -.083607 | -.275987 | .438149 |
| 5 11/2 | .330859 | .688220 | .925574 | 1.000000 | .925359 | .682263 | -.403743 |
| 7 13/2 | .245354 | .164774 | .081480 | .000000 | -.159226 | -.421537 | .561678 |
| 7 15/2 | .366866 | .323536 | .181836 | .000000 | -.178626 | -.303194 | .349615 |

**K = 3/2-[741]**

| | | | | | | | |
|---|---|---|---|---|---|---|---|
| E | 7.749823 | 7.783581 | 7.768100 | 7.684800 | 7.541741 | 7.365543 | 7.190500 |
| L J | | | | | | | |
| 1 3/2 | -.061230 | .034909 | .004945 | .000000 | .002636 | .039613 | .102130 |
| 3 5/2 | .327090 | .211199 | .056096 | .000000 | .035824 | .061877 | .028919 |
| 3 7/2 | .256106 | .082235 | .016565 | .000000 | .041102 | .218164 | .395190 |
| 5 9/2 | -.538194 | -.472782 | -.262559 | .000000 | .229112 | .312332 | .270272 |
| 5 11/2 | -.244058 | -.163832 | -.072027 | .000000 | .171154 | .436876 | .531000 |
| 7 13/2 | .644167 | .829406 | .960056 | 1.000000 | .956561 | .800157 | .628625 |
| 7 15/2 | -.243935 | -.095579 | -.026864 | .000000 | -.015009 | -.135440 | -.287021 |

| β = | | -0.3 | -0.2 | -0.1 | 0.0 | 0.1 | 0.2 | 0.3 |
|---|---|---|---|---|---|---|---|---|

**K = 3/2-[732]**

| | | -0.3 | -0.2 | -0.1 | 0.0 | 0.1 | 0.2 | 0.3 |
|---|---|---|---|---|---|---|---|---|
| E | | 7.997871 | 8.025661 | 8.110296 | 8.089600 | 7.980147 | 7.847280 | 7.736036 |
| L | J | | | | | | | |
| 1 | 3/2 | -.605984 | .733123 | -.585576 | .000000 | .230470 | -.250680 | -.239866 |
| 3 | 5/2 | -.338121 | .328601 | -.148332 | .000000 | -.307968 | .465403 | .516809 |
| 3 | 7/2 | -.087285 | -.231497 | .703268 | 1.000000 | .690736 | -.403895 | -.244317 |
| 5 | 9/2 | .018459 | -.171290 | .247882 | .000000 | -.539249 | .564255 | .498698 |
| 5 | 11/2 | .501603 | -.420584 | .265325 | .000000 | -.243988 | .345828 | .392779 |
| 7 | 13/2 | .446589 | -.285535 | .088090 | .000000 | .154462 | -.336132 | -.440211 |
| 7 | 15/2 | .243826 | -.114889 | .030152 | .000000 | .027282 | -.080389 | -.137554 |

**K = 3/2-[721]**

| | | -0.3 | -0.2 | -0.1 | 0.0 | 0.1 | 0.2 | 0.3 |
|---|---|---|---|---|---|---|---|---|
| E | | 8.346304 | 8.258639 | 8.207618 | 8.149000 | 8.061455 | 8.028157 | 8.020887 |
| L | J | | | | | | | |
| 1 | 3/2 | .308426 | .234138 | .031898 | .000000 | .146784 | .274087 | .334663 |
| 3 | 5/2 | -.566203 | -.571199 | -.475570 | .000000 | .185865 | .132643 | .095069 |
| 3 | 7/2 | -.164618 | -.305206 | -.321846 | .000000 | .595603 | .668873 | .628266 |
| 5 | 9/2 | .372163 | .515939 | .771537 | 1.000000 | .731839 | .548545 | .483073 |
| 5 | 11/2 | -.403426 | -.321093 | -.138724 | .000000 | -.141542 | -.313988 | -.410442 |
| 7 | 13/2 | .480989 | .387217 | .233295 | .000000 | -.182724 | -.239683 | -.269398 |
| 7 | 15/2 | -.156778 | -.080805 | -.017026 | .000000 | .012457 | .054339 | .099226 |

**K = 3/2-[701]**

| | | -0.3 | -0.2 | -0.1 | 0.0 | 0.1 | 0.2 | 0.3 |
|---|---|---|---|---|---|---|---|---|
| E | | 8.696780 | 8.495996 | 8.356187 | 8.406600 | 8.591492 | 8.791745 | 9.000036 |
| L | J | | | | | | | |
| 1 | 3/2 | .321412 | .443848 | .721818 | 1.000000 | .851479 | .851928 | .845074 |
| 3 | 5/2 | .238559 | .239053 | .3 6187 | .000000 | .473020 | .419091 | .393361 |
| 3 | 7/2 | .622834 | .636675 | .488921 | .000000 | -.176896 | -.258359 | -.302299 |
| 5 | 9/2 | .492422 | .476601 | .359078 | .000000 | -.138392 | -.167276 | -.178540 |
| 5 | 11/2 | .372312 | .274551 | .106395 | .000000 | .022664 | .053435 | .078174 |
| 7 | 13/2 | .249357 | .188789 | .076405 | .000000 | .016608 | .030965 | .040168 |
| 7 | 15/2 | .092666 | .048267 | .009469 | .000000 | -.001476 | -.005980 | -.011415 |

**K = 3/2-[712]**

| | | -0.3 | -0.2 | -0.1 | 0.0 | 0.1 | 0.2 | 0.3 |
|---|---|---|---|---|---|---|---|---|
| E | | 9.150215 | 8.853551 | 8.592937 | 8.439600 | 8.430436 | 8.500420 | 8.594184 |
| L | J | | | | | | | |
| 1 | 3/2 | -.316560 | -.336937 | -.361013 | .000000 | -.445833 | -.352881 | -.297367 |
| 3 | 5/2 | .605390 | .680548 | .809184 | 1.000000 | .802890 | .755443 | .721600 |
| 3 | 7/2 | -.426414 | -.371795 | -.251237 | .000000 | .240112 | .324532 | .360673 |
| 5 | 9/2 | .525706 | .498249 | .381284 | .000000 | -.308158 | -.422552 | -.467597 |
| 5 | 11/2 | -.204362 | -.137556 | -.053729 | .000000 | -.046099 | -.106634 | -.153483 |
| 7 | 13/2 | .177224 | .132067 | .059026 | .000000 | .042959 | .096389 | .132890 |
| 7 | 15/2 | -.043868 | -.021921 | -.004725 | .000000 | .003809 | .015734 | .030282 |

**K = 1/2-[770]**

| | | -0.3 | -0.2 | -0.1 | 0.0 | 0.1 | 0.2 | 0.3 |
|---|---|---|---|---|---|---|---|---|
| E | | 7.106570 | 7.099426 | 7.042204 | 6.934800 | 6.777359 | 6.570322 | 6.314898 |
| a | | -7.280311 | -7.689804 | -7.923812 | -8.000000 | -7.924127 | -7.693081 | -7.298259 |
| L | J | | | | | | | |
| 1 | 1/2 | -.031014 | -.006495 | -.000409 | .000000 | -.000357 | -.004943 | -.020380 |
| 1 | 3/2 | -.082058 | -.027039 | -.0 3572 | .000000 | .003435 | .024932 | .072110 |
| 3 | 5/2 | .014343 | .003557 | .000436 | .000000 | -.000648 | -.006356 | -.024183 |

| | β = | −0.3 | −0.2 | −0.1 | 0.0 | 0.1 | 0.2 | 0.3 |
|---|---|---|---|---|---|---|---|---|
| 3 | 7/2 | .215555 | .105791 | .027956 | .000000 | .027643 | .103242 | .207063 |
| 5 | 9/2 | −.010462 | −.004561 | −.001337 | .000000 | −.001920 | −.009320 | −.025609 |
| 5 | 11/2 | −.489859 | −.360270 | −.190672 | .000000 | .190010 | .357824 | .485541 |
| 7 | 13/2 | .008813 | .006646 | .004051 | .000000 | −.005681 | −.013374 | −.024281 |
| 7 | 15/2 | .839932 | .926371 | .981240 | 1.000000 | .981368 | .927551 | .844943 |

**K = 1/2 - [761]**

| | | | | | | | | |
|---|---|---|---|---|---|---|---|---|
| E | | 7.674886 | 7.726011 | 7.696666 | 7.599000 | 7.434707 | 7.204292 | 6.905641 |
| a | | −2.320425 | −5.234631 | −5.821672 | −6.000000 | −5.787855 | −4.503232 | −1.336666 |
| L | J | | | | | | | |
| 1 | 1/2 | .368187 | .118970 | .015007 | .000000 | −.012667 | −.082762 | −.184486 |
| 1 | 3/2 | .401177 | .263047 | .077019 | .000000 | .072402 | .229481 | .329143 |
| 3 | 5/2 | −.266865 | −.053607 | −.006901 | .000000 | −.018202 | −.119193 | −.290741 |
| 3 | 7/2 | −.507651 | −.545205 | −.326845 | .000000 | .322123 | .517934 | .488191 |
| 5 | 9/2 | .243047 | .026028 | .001723 | .000000 | −.032883 | −.160504 | −.356666 |
| 5 | 11/2 | .341443 | .704767 | .922268 | 1.000000 | .922488 | .692234 | .359701 |
| 7 | 13/2 | −.212687 | .001006 | .027904 | .000000 | −.054482 | −.199428 | −.356963 |
| 7 | 15/2 | .392018 | .345188 | .188701 | .000000 | −.188333 | −.336609 | −.388266 |

**K = 1/2 - [750]**

| | | | | | | | | |
|---|---|---|---|---|---|---|---|---|
| E | | 7.758574 | 7.801450 | 7.780885 | 7.684800 | 7.517569 | 7.285415 | 7.022945 |
| a | | 3.219062 | 6.235471 | 6.830292 | 7.000000 | 6.794867 | 5.497283 | 2.258660 |
| L | J | | | | | | | |
| 1 | 1/2 | .120895 | −.111427 | −.014312 | .000000 | .009948 | .041601 | .019133 |
| 1 | 3/2 | −.149095 | −.013882 | −.001714 | .000000 | −.000934 | .027147 | .158997 |
| 3 | 5/2 | −.308529 | .225143 | .061077 | .000000 | .057428 | .172140 | .190929 |
| 3 | 7/2 | .297617 | −.016820 | .006849 | .000000 | .015230 | .133024 | .354333 |
| 5 | 9/2 | .514402 | −.479464 | −.268908 | .000000 | .264531 | .442166 | .449013 |
| 5 | 11/2 | −.260497 | .040978 | −.023440 | .000000 | .060938 | .227384 | .346244 |
| 7 | 13/2 | −.631248 | .839523 | .960766 | 1.000000 | .960543 | .834086 | .650115 |
| 7 | 15/2 | −.220107 | .007423 | −.009122 | .000000 | −.006105 | −.085384 | −.261152 |

**K = 1/2 - [741]**

| | | | | | | | | |
|---|---|---|---|---|---|---|---|---|
| E | | 8.058582 | 8.099880 | 8.161935 | 8.089600 | 7.906721 | 7.647684 | 7.416801 |
| a | | .852338 | .390231 | −3.197373 | −4.000000 | −2.059245 | .326631 | .829328 |
| L | J | | | | | | | |
| 1 | 1/2 | .589101 | .597887 | −.250174 | .000000 | −.175747 | −.335924 | .374135 |
| 1 | 3/2 | .240158 | .406903 | −.494451 | .000000 | .431452 | .442384 | −.362177 |
| 3 | 5/2 | −.345735 | −.421514 | .133796 | .000000 | −.274812 | −.482067 | .491341 |
| 3 | 7/2 | .125167 | −.163577 | .755846 | 1.000000 | .713154 | .280088 | −.033793 |
| 5 | 9/2 | −.044511 | .171771 | −.080462 | .000000 | −.318728 | −.306040 | .162695 |
| 5 | 11/2 | −.405975 | −.365669 | .307943 | .000000 | −.288425 | −.400025 | .440437 |
| 7 | 13/2 | .493288 | .312746 | −.033165 | .000000 | .113653 | .339520 | −.475414 |
| 7 | 15/2 | −.223504 | −.107241 | .036368 | .000000 | .034035 | .107979 | −.199548 |

**K = 1/2 - [730]**

| | | | | | | | | |
|---|---|---|---|---|---|---|---|---|
| E | | 8.356005 | 8.258824 | 8.213460 | 8.149000 | 7.977871 | 7.825103 | 7.708571 |
| a | | .037998 | .414427 | 4.069774 | 5.000000 | 2.988555 | .543279 | −.034748 |
| L | J | | | | | | | |
| 1 | 1/2 | −.031189 | .050304 | .223344 | .000000 | .077273 | .015125 | −.033718 |
| 1 | 3/2 | −.495645 | −.449447 | −.063067 | .000000 | .153889 | .387591 | .460533 |
| 3 | 5/2 | −.411673 | −.426146 | −.485598 | .000000 | .383383 | .362275 | .339733 |
| 3 | 7/2 | .225455 | .395748 | .189951 | .000000 | .402151 | .463713 | .347946 |

| l | J | β = −0.3 | −0.2 | −0.1 | 0.0 | 0.1 | 0.2 | 0.3 |
|---|---|---|---|---|---|---|---|---|
| 5 | 9/2 | .319894 | .433277 | .779446 | 1.000000 | .768570 | .514540 | .416063 |
| 5 | 11/2 | .468798 | .373180 | .055123 | .000000 | −.127561 | −.350451 | −.452416 |
| 7 | 13/2 | .425665 | .349911 | .252286 | .000000 | −.233434 | −.331529 | −.388916 |
| 7 | 15/2 | .172524 | .088431 | .005399 | .000000 | .013265 | .076115 | .145750 |

K = 1/2−[710]

| | | −0.3 | −0.2 | −0.1 | 0.0 | 0.1 | 0.2 | 0.3 |
|---|---|---|---|---|---|---|---|---|
| E | | 9.164536 | 8.858399 | 8.587508 | 8.406600 | 8.415417 | 8.483102 | 8.576013 |
| a | | −.019023 | .013091 | .118094 | −2.000000 | .253518 | −.022569 | −.173477 |
| l | J | | | | | | | |
| 1 | 1/2 | .044195 | .013217 | −.051553 | .000000 | .033824 | −.018740 | −.050185 |
| 1 | 3/2 | .491041 | .568422 | .674524 | 1.000000 | .683097 | .633583 | .588293 |
| 3 | 5/2 | .469938 | .496828 | .567258 | .000000 | .610458 | .537379 | .504398 |
| 3 | 7/2 | .509561 | .468328 | .333888 | .000000 | −.280381 | −.406457 | −.460556 |
| 5 | 9/2 | .445498 | .414069 | .319659 | .000000 | −.277243 | −.349779 | −.374838 |
| 5 | 11/2 | .226636 | .158429 | .063797 | .000000 | .048598 | .119730 | .174493 |
| 7 | 13/2 | .156199 | .115832 | .053070 | .000000 | .041506 | .085310 | .113386 |
| 7 | 15/2 | .046897 | .024115 | .005282 | .000000 | −.003790 | −.016680 | −.032491 |

K = 1/2−[721]

| | | −0.3 | −0.2 | −0.1 | 0.0 | 0.1 | 0.2 | 0.3 |
|---|---|---|---|---|---|---|---|---|
| E | | 9.646522 | 9.231478 | 8.835284 | 8.439600 | 8.246229 | 8.179264 | 8.160059 |
| a | | .699437 | .879922 | 1.077440 | 3.000000 | .722100 | .924452 | .908600 |
| l | J | | | | | | | |
| 1 | 1/2 | −.483557 | −.526575 | .621357 | .000000 | −.502700 | −.508514 | −.496398 |
| 1 | 3/2 | .496800 | .489045 | −.453494 | .000000 | .461206 | .240886 | .122247 |
| 3 | 5/2 | −.565238 | −.583169 | .588051 | 1.000000 | −.484915 | −.334670 | −.265185 |
| 3 | 7/2 | .313672 | .260197 | −.163436 | .000000 | −.370895 | −.469904 | −.471424 |
| 5 | 9/2 | −.291587 | −.261130 | .185710 | .000000 | .386682 | .524790 | .548066 |
| 5 | 11/2 | .104240 | .068649 | −.026415 | .000000 | .084829 | .206526 | .287931 |
| 7 | 13/2 | −.072618 | −.052762 | .023544 | .000000 | −.071395 | −.176667 | −.240556 |
| 7 | 15/2 | .017758 | .008898 | −.001974 | .000000 | −.007752 | −.036121 | −.070227 |

K = 1/2−[701]

| | | −0.3 | −0.2 | −0.1 | 0.0 | 0.1 | 0.2 | 0.3 |
|---|---|---|---|---|---|---|---|---|
| E | | 8.743615 | 8.550724 | 8.425154 | 8.556600 | 8.701031 | 8.898626 | 9.105782 |
| a | | .810924 | .991294 | .847257 | 1.000000 | 1.012188 | .927238 | .846562 |
| l | J | | | | | | | |
| 1 | 1/2 | −.514846 | .579597 | .705945 | 1.000000 | .842040 | .787009 | .758379 |
| 1 | 3/2 | .137195 | −.006356 | .291467 | .000000 | .324931 | .374962 | .396092 |
| 3 | 5/2 | −.063593 | −.015022 | −.273732 | .000000 | −.407510 | −.437845 | −.443684 |
| 3 | 7/2 | −.427447 | .469968 | .383100 | .000000 | −.093544 | −.154577 | −.191523 |
| 5 | 9/2 | .537033 | −.559052 | −.420716 | .000000 | .101312 | .149929 | .173387 |
| 5 | 11/2 | −.358858 | .262754 | .099408 | .000000 | .013436 | .035308 | .054430 |
| 7 | 13/2 | .319400 | −.242296 | −.089530 | .000000 | −.011317 | −.025974 | −.036597 |
| 7 | 15/2 | −.100776 | .051437 | .009675 | .000000 | −.000934 | −.004168 | −.008350 |

# BIBLIOGRAPHY

## I. Books


1. L. C. Biedenharn and P. J. Brussaard, "Coulomb Excitation," Oxford Univ. Press (Clarendon), London and New York, 1965.
2. S. Bjørnholm, "Nuclear Excitations in Even Isotopes of the Heaviest Elements," Munksgaard, Copenhagen, 1965.
3. F. Bowman, "Introduction to Elliptic Functions with Applications," English Univ. Press, London, 1953; *see also* L. M. Milne-Thomson, "Jacobian Elliptic Function Tables," Dover, New York, 1950.
4. H. B. G. Casimir, "On the Interaction between Atomic Nuclei and Electrons," 2nd ed., Freeman, San Francisco, 1963.
5. E. U. Condon and G. H. Shortley, "Theory of Atomic Spectra," 2nd ed., Cambridge Univ. Press, London and New York, 1957.
6. A. de-Shalit and I. Talmi, "Nuclear Shell Theory," Academic Press, New York, 1963.
7. P. A. M. Dirac, "Principles of Quantum Mechanics," 4th ed., Oxford Univ. Press (Clarendon), London and New York, 1958.
8. H. Goldstein, "Classical Mechanics," Addison-Wesley, Reading, Massachusetts, 1953.
9. E. M. Henley and W. Thirring, "Elementary Quantum Field Theory," McGraw-Hill, New York, 1962.
10. Z. Kopal, "Numerical Analysis," 2nd ed., Wiley, New York, 1961.
11. L. D. Landau and E. M. Lifschitz, "Quantum Mechanics—Nonrelativistic Theory," Pergamon Press, Oxford, 1958.
12. J. S. Levinger, "Nuclear Photodisintegration," Oxford Univ. Press, London and New York, 1960.
13. R. A. Lyttleton, "The Stability of Rotating Liquid Masses," Cambridge Univ. Press, London and New York, 1953.
14. M. A. Preston, "Physics of the Nucleus," Addison-Wesley, Reading, Massachusetts, 1962.
15. M. E. Rose, "Elementary Theory of Angular Momentum," Wiley, New York, 1957.
16. L. I. Schiff, "Quantum Mechanics," 2nd. ed., Chap. XIII, McGraw-Hill, New York, 1955.

17. J. W. Strutt (Lord Rayleigh), "The Theory of Sound," Vol. II, p. 264 ff, Dover, New York, 1945.
18. C. H. Townes and A. L. Schawlow, "Microwave Spectroscopy," McGraw-Hill, New York, 1955.
19. E. T. Whittaker, "A Treatise on the Analytical Dynamics of Particles and Rigid Bodies, 4th ed., Cambridge Univ. Press, London and New York, 1927.
20. E. T. Whittaker and G. N. Watson, "Modern Analysis," 4th Ed., Cambridge Univ. Press, London and New York, 1927.
21. E. P. Wigner, "Group Theory," Academic Press, New York, 1959.
22. C. S. Wu and S. A. Moszkowski, "Beta Decay," Wiley (Interscience), New York, 1966.

## II. Review Articles

23. K. Adler, A. Bohr, T. Huus, B. Mottelson, and A. Winther, *Rev. Mod. Phys.* **28**, 432 (1956).
24. W. C. Barber, *Ann. Rev. Nucl. Sci.* **12**, 1 (1962).
25. H. A. Bethe, *Rev. Mod. Phys.* **9**, 69 (1937).
26. L. A. Borisoglebskii, *Soviet Phys. Usp.* (*English Transl.*) **6**, 715 (1964).
27. M. Danos and E. G. Fuller, *Ann. Rev. Nucl. Sci.* **15**, 29 (1965).
28. J. P. Davidson, *Rev. Mod. Phys.* **37**, 105 (1965).
29. T. de Forest, Jr. and J. D. Walecka, *Advan. Phys.* **15**, 1 (1966).
30. C. J. Gallagher, Jr. and V. G. Soloviev, *Kgl. Danske Videnskab. Selskab Mat. Fys. Skrifter* **2**, 2 (1962).
31. R. Hofstadter, *Rev. Mod. Phys.* **28**, 214 (1956).
32. H. J. Mang, *Ann. Rev. Nucl. Sci.* **14**, 1 (1964).
33. S. A. Moszkowski, *in* "Handbuch der Physik," (S. Flugge, ed.), Vol. 39, p. 411, Springer, Berlin, 1952.
34. B. R. Mottelson and S. G. Nilsson, *Kgl. Danske Videnskab. Selskab Mat. Fys. Skrifter* **1**, 8 (1959).
35. W. Pauli, *in* "Handbuch der Physik," (S. Flugge, ed.), Vol. 5, Part 1, Springer, Berlin, 1958.
36. C. F. Perdrisat, *Rev. Mod. Phys.* **38**, 41 (1966).
37. I. Perlman and J. O. Rasmussen, *in* "Handbuch der Physik," (S. Flugge, ed.), Vol. 42, p. 109, Springer, Berlin, 1957.
38. J. Serrin, *in* "Encyclopedia of Physics," (S. Flugge, ed.), Vol. 8, Part. 1, p. 125, Springer, Berlin, 1959.
39. K. Way, *Natl. Bur. Std.* (*U. S.*), *Circ. 499* (1950).
40. A. H. Wapstra, *Nucl. Data* **1A**, 21 (1965).
41. J. A. Wheeler, *Rev. Mod. Phys.* **21**, 133 (1949).
42. D. H. Wilkinson, *Ann. Rev. Nucl. Sci.* **9**, 1 (1959).

## III. Articles


43. H. L. Acker, H. Marschall, G. Barkenstoss, and D. Quitman, *Nucl. Phys.* **62**, 477 (1965).
44. E. Ambler, E. G. Fuller, and H. Marshak, *Phys. Rev.* **B138**, 117 (1965).
45. M. Yu. Balats et al., *Soviet Phys. JETP* (*English Transl.*) **22**, 4 (1966).
46. G. C. Baldwin and G. S. Klaiber, *Phys. Rev.* **71**, 3 (1947); *ibid.* **73**, 1156 (1948).
47. M. Baranger and K. Kumar, *Nucl. Phys.* **62**, 113 (1965).
48. S. A. Baranov, B. M. Kulakov, and S. N. Belenky, *Nucl. Phys.* **41**, 95 (1963).
49. S. T. Belyaev, *Kgl. Danske Videnskab. Selskab Mat. Fys. Medd.* **31**, 11 (1959).
50. R. Beringer and W.J. Knox, *Phys. Rev.* **121**, 1195 (1961).
51. B. L. Birbrair, L. K. Peker, and L. A. Sliv, *Soviet Phys. JETP* (*English Transl.*) **9**, 566 (1959).
52. S. Bjørnholm, M. Lederer, F. Asaro, and I. Perlman, *Phys. Rev.* **130**, 2000 (1963).
53. R. L. Bramblett, J. T. Caldwell, G. F. Auchampaugh, and S. C. Fultz, *Phys. Rev.* **B133**, 869 (1964).
54. S. Brenner, *Phil. Mag.* **47**, 429 (1956).
55. A. Bohr, *Phys. Rev.* **81**, 134 (1951).
56. A. Bohr, *Kgl. Danske Videnskab. Selskab Mat. Fys. Medd.* **26**, 14 (1952).
57. A. Bohr and B. R. Mottelson, *Kgl. Danske Videnskab Selskab Mat. Fys. Medd.* **27**, 16 (1953).
58. N. Bohr, *Trans. Roy. Soc.* (London) **A209**, 281 (1909).
59. N. Bohr and F. Kalcker, *Kgl. Danske Videnskab. Selskab Mat. Fys. Medd.* **14**, 10 (1937).
60. N. Bohr and J. A. Wheeler, *Phys. Rev.* **56**, 426 (1939).
61. W. Y. Chang. *Rev. Mod. Phys.* **21**, 166 (1949).
62. B. E. Chi and J. P. Davidson, *Phys. Rev.* **131**, 366 (1963).
63. R. F. Christy, *Phys. Rev.* **98**, 1025 (1955).
64. E. L. Church and J. Weneser, *Phys. Rev.* **103**, 1035 (1956).
65. E. U. Condon and R. W. Gurney, *Nature* **122**, 439 (1928); *Phys. Rev.* **33**, 127 (1929).
66. L. N. Cooper and E. M. Henley, *Phys. Rev.* **92**, 801 (1953).
67. M. Danos, *Nucl. Phys.* **5**, 23 (1958).
68. M. Danos and W. Greiner, *Phys. Rev.* **B134**, 284 (1964).
69. M. Danos and W. Greiner, *Phys. Letters* **8**, 113 (1964).
70. M. Danos, W. Greiner, and C. B. Kohr, *Phys. Rev.* **B138**, 1055 (1965).
71. J. P. Davidson and E. Feenberg, *Phys. Rev.* **89**, 856 (1953).
72. J. P. Davidson, *Nucl. Phys.* **33**, 664 (1962).
73. J. P. Davidson and M. G. Davidson, *Phys. Rev.* **B138**, 316 (1965).
74. J. P. Davidson, *in* "Nuclear Spin-Parity Assignments," (N. B. Grove, ed.), p. 446 ff, Academic Press, New York, 1966.
75. J. P. Davidson, *Nucl. Phys.* **86**, 561 (1966).
76. M. G. Davidson, *Nucl. Phys.* **69**, 455 (1965).
77. M. G. Davidson, *Phys. Letters* **22**, 596 (1966).
78. A. S. Davydov and G. F. Filippov, *Nucl. Phys.* **8**, 237 (1958).

79. A. S. Davydov and A. A. Chaban, *Nucl. Phys.* **20**, 499 (1960).
80. A. S. Davydov, *Nucl. Phys.* **24**, 682 (1961).
81. A. S. Davydov, V. S. Rostovsky, and A. A. Chaban, *Nucl. Phys.* **27**, 134 (1961); A. S. Davydov and V. S. Rostovsky, *Nucl. Phys.* **60**, 529 (1964).
82. R. M. Diamond, B. Elbek, and F. S. Stephens, *Nucl. Phys.* **43**, 560 (1963).
83. R. M. Diamond, F. S. Stephens, and W. J. Swiatecki, *Phys. Letters* **11**, 315, (1964).
84. J. Diaz, S. N. Kaplan, B. MacDonald, and R. V. Pyle, *Phys. Rev. Letters* **3**, 234 (1959).
85. R. D. Ehrlich et al., *Phys. Rev. Letters* **13**, 550 (1964); S. Raboy et al., *Nucl. Phys.* **73**, 353 (1965).
86. W. Elsasser, *J. Phys. Radium* **5**, 625 (1934).
87. A. Faessler and W. Greiner, *Z. Physik* **168**, 425 (1962); *ibid.* **170**, 105 (1962); *ibid.* **177**, 190 (1964).
88. E. Feenberg, *Phys. Rev.* **55**, 504 (1939).
89. E. Feenberg and K. C. Hammack, *Phys. Rev.* **81**, 285 (1951).
90. V. L. Fitch and J. Rainwater, *Phys. Rev.* **92**, 789 (1953).
91. L. L. Foldy and F. J. Milford, *Phys. Rev.* **80**, 751 (1950).
92. K. W. Ford and J. G. Wills, "Calculated Properties of Mu-Mesonic Atoms," Los Alamos Scientific Publication LAMS-2387, Los Alamos, New Mexico, 1960.
93. J. Frenkel, *Phys. Rev.* **55**, 987 (1939); *J. Phys.* **1**, 125 (1939).
94. P. O. Froman, *Kgl. Danske Videnskab. Selskab Mat. Fys. Skrifter* **1**, 3 (1957).
95. E. G. Fuller and M. S. Weiss, *Phys. Rev.* **112**, 560 (1958).
96. E. G. Fuller and E. Hayward, *Nucl. Phys.* **30**, 613 (1962).
97. C. J. Gallagher, Jr., *Nucl. Phys.* **16**, 215 (1960).
98. C. J. Gallagher, Jr. and S. A. Moszkowski, *Phys. Rev.* **111**, 1282 (1958).
99. G. Gamow, *Nature* **126**, 387 (1930).
100. G. Gamow, *Z. Physik* **51**, 204 (1958).
101. H. Geiger and J. M. Nuttall, *Phil. Mag.* **22**, 613 (1911).
102. A. Ghiorso, S. G. Thompson, G. H. Higgins, B. G. Harvey, and G. T. Seaborg, *Phys. Rev.* **95**, 293 (1954).
103. M. Goldhaber and E. Teller, *Phys. Rev.* **74**, 1046 (1948).
104. W. Gordy, *Phys. Rev.* **76**, 139 (1949).
105. H. B. Greenstein, *Phys. Rev.* **144**, 902 (1966).
106. K. T. Hecht and G. R. Satchler, *Nucl. Phys.* **32**, 286 (1962).
107. E. K. Hege, *Phys. Rev.* (in press.).
108. D. L. Hill and J. A. Wheeler, *Phys. Rev.* **89**, 1102 (1953).
109. W. M. Hooke, *Phys. Rev.* **115**, 453 (1959).
110. E. V. Inopin, *Soviet Phys. JETP* **11**, 714 (1960); K. Okamoto, *Progr. Theoret. Phys. (Kyoto)* **28**, 1073 (1962).
111. B. A. Jacobsohn, *Phys. Rev.* **96**, 1637 (1954).
112. Z. Jankovic, *Nuovo Cimento* **14**, 1174 (1959).
113. H. W. Kendall and J. Oeser, *Phys. Rev.* **130**, 245 (1963); *see also* H. Crannell et al., *Phys. Rev.* **123**, 923 (1961).
114. A. K. Kerman, *Kgl. Danske Videnskab. Selskab Mat. Fys. Medd.* **30**, 15 (1956).
115. G. W. King, R. M. Hainer, and P. C. Cross, *J. Chem. Phys.* **11**, 27 (1943).

116. K. Kjallquist, *Nucl. Phys.* **9**, 163 (1958/59); K. Harada, *Phys. Letters* **10**, 80 (1964).
117. A. M. Lane and E. D. Pendlebury, *Nucl. Phys.* **15**, 39 (1960).
118. M. LeBellac, *Nucl. Phys.* **40**, 645 (1963) for a theoretical discussion; R. D. Ehrlich et al., *Phys. Rev. Letters* **16**, 425 (1966) and T. T. Bardin et al., *Phys. Rev. Letters* **16**, 429 (1966) for experimental verification.
119. C. M. Lederer, "The Structure of Heavy Nuclei: A Study of Very Weak Alpha Branching", UCRL-11028, Berkeley, California, 1963.
120. C. M. Lederer et al., *Nucl. Phys.* **84**, 481 (1966).
121. D. P. Leper, *Nucl. Phys.* **50**, 234 (1964).
122. J. LeTourneux, *Kgl. Danske Videnskab. Selskab Mat. Fys. Medd.* **34**, 11 (1965).
123. P. O. Lipas and J. P. Davidson, *Nucl. Phys.* **26**, 80 (1961).
124. P. O. Lipas, *Nucl. Phys.* **39**, 468 (1962).
125. A. E. Litherland, E. B. Paul, G. A. Bartholomew, and H. E. Gove, *Phys. Rev.* **102**, 208 (1956).
126. R. McLaurin, *Trans. Cambridge Phil. Soc.* **17**, 41 (1898).
127. C. A. Mallman and A. K. Kerman, *Nucl. Phys.* **16**, 105 (1960).
128. C. A. Mallman, *Nucl. Phys.* **24**, 535 (1961).
129. C. Marty, *Nucl. Phys.* **1**, 85 (1956); *ibid.* **3**, 193 (1957).
130. F. C. Michel, *Phys. Rev.* **B133**, 239 (1964).
131. J. W. Mihelich, G. Scharff-Goldhaber, and M. McKeown, *Phys. Rev.* **94**, 794 (1954).
132. R. B. Moore and W. White, *Can. J. Phys.* **38**, 1149 (1960).
133. T. D. Newton, *Can. J. Phys.* **38**, 700 (1960).
134. S. G. Nilsson, *Kgl. Danske Videnskab. Selskab Mat. Fys. Medd.* **29**, 16 (1955).
135. S. G. Nilsson and O. Prior, *Kgl. Danske Videnskab. Selskab Mat. Fys. Medd.* **32**, 16 (1961).
136. L. W. Nordheim, *Phys. Rev.* **78**, 294 (1950); *Rev. Mod. Phys.* **23**, 322 (1951).
137. I. Perlman, A. Ghiorso, and G. T. Seaborg, *Phys. Rev.* **77**, 76 (1950).
138. A. K. Rafiqullah, *Phys. Rev.* **127**, 905 (1962).
139. J. Rainwater, *Phys. Rev.* **79**, 432 (1950).
140. J. O. Rasmussen, *Nucl. Phys.* **19**, 85 (1960).
141. A. S. Reiner, *Nucl. Phys.* **27**, 115 (1961).
142. M. E. Rose and R. K. Osborn, *Phys. Rev.* **93**, 1326 (1954).
143. S. Rosenblum, *Compt. Rend.* **188**, 1401 (1929); *ibid.* **190**, 1124 (1930).
144. J. E. Russell, *Phys. Rev.* **127**, 245 (1962).
145. E. Rutherford, *Phil. Mag.* **21**, 669 (1911).
146. G. Scharff-Goldhaber and J. Weneser, *Phys. Rev.* **98**, 212 (1955).
147. V. G. Soloviev, P. Vogel, and A. A. Korneichuk, *Izv. Akad. Nauk SSSR, Ser. Fiz*, **28**, 1599 (1964).
148. M. B. Stearns and M. Stearns, *Phys. Rev.* **105**, 1573 (1957); J. L. Lathrop et al., *Phys. Rev. Letters* **7**, 147 (1961); and M. B. Stearns, private communication.
149. H. Steinwedel and J. H. D. Jensen, *Z. Naturforsch.* **5a**, 413 (1950); H. Steinwedel, J. H. D. Jensen, and P. Jensen, *Phys. Rev.* **79**, 1019 (1950).
150. F. S. Stephens, Jr., R. M. Diamond, and I. Perlman, *Phys. Rev. Letters* **3**, 435 (1959).

151. J. W. Strutt (Lord Rayleigh), *Proc. Roy. Soc.* (London) **29**, 71 (1879).
152. J. Thibaud, *Compt. Rend.* **191**, 656 (1930).
153. H. A. Tolhoek and J. A. M. Cox, *Physica* **19**, 101 (1953).
154. C. H. Townes, H. M. Foley, and W. Low, *Phys. Rev.* **76**, 1415 (1949).
155. M. N. Vergnes and J. O. Rasmussen, *Nucl. Phys.* **62**, 233 (1965); *see also* Ref. 36.
156. C. F. von Weizsäcker, *Naturwiss.* **27**, 133 (1939).
157. J. A. Wheeler, *Phys. Rev.* **92**, 812 (1953).
158. L. Wilets, *Kgl. Danske Videnskab. Selskab Mat. Fys. Medd.* **29**, 3 (1954).
159. S. A. Williams, *Phys. Rev.* **125**, 340 (1962).
160. S. A. Williams and J. P. Davidson, *Can. J. Phys.* **40**, 1423 (1962).
161. S. A. Williams, *Nucl. Phys.* **63**, 581 (1965).
162. J. H. Zablotney, Master's thesis, Rensselaer Polytechnic Inst., Troy, New York, 1967. (Unpublished.)
163. D. F. Zaretski and V. M. Novikov, *Nucl. Phys.* **28**, 177 (1961).
164. D. F. Zaretski and V. M. Novikov, *Soviet Phys. JETP* **14**, 157 (1962); *see also* V. M. Novikov, *Soviet Phys. JETP* **14**, 198 (1962).

# Author Index

Numbers in parentheses are reference numbers and indicate that an author's work is referred to, although his name is not cited in the text. Numbers in italics show the page on which the complete reference is listed.

## S



## T



## V



## W



## Z

# Subject Index

## A



## B



## C



## D

## E



## F



## G



## H



## I



## J



## K



## L



## M